CAMBRIDGE LIBRARY

Books of enduring scholarly

Perspectives from the Royal Asiatic Society

A long-standing European fascination with Asia, from the Middle East to China and Japan, came more sharply into focus during the early modern period, as voyages of exploration gave rise to commercial enterprises such as the East India companies, and their attendant colonial activities. This series is a collaborative venture between the Cambridge Library Collection and the Royal Asiatic Society of Great Britain and Ireland, founded in 1823. The series reissues works from the Royal Asiatic Society's extensive library of rare books and sponsored publications that shed light on eighteenth- and nineteenth-century European responses to the cultures of the Middle East and Asia. The selection covers Asian languages, literature, religions, philosophy, historiography, law, mathematics and science, as studied and translated by Europeans and presented for Western readers.

A Synopsis of Science

James Robert Ballantyne (1813–64) taught oriental languages in India for sixteen years, producing grammars of Hindi, Sanskrit and Persian, along with translations of Hindu philosophy. In 1859, for the use of Christian missionaries, he prepared a guide to Hinduism, in English and Sanskrit (also reissued in this series). Published in two volumes in 1852, *Synopsis of Science* was intended to introduce his Indian pupils to Western science by using the framework of Hindu Nyaya philosophy, which was familiar to them and which Ballantyne greatly respected. Volume 1 proceeds through a series of aphorisms exemplifying Western logic, forms of argument, the workings of the body and the senses, basic astronomy, geography, laws of Newtonian motion, the action of liquids, pneumatics, acoustics and optics. The second part of the volume is a Sanskrit translation. Overall, the work serves as an excellent primary source on the educational aspects of British imperialism.

Cambridge University Press has long been a pioneer in the reissuing of out-of-print titles from its own backlist, producing digital reprints of books that are still sought after by scholars and students but could not be reprinted economically using traditional technology. The Cambridge Library Collection extends this activity to a wider range of books which are still of importance to researchers and professionals, either for the source material they contain, or as landmarks in the history of their academic discipline.

Drawing from the world-renowned collections in the Cambridge University Library and other partner libraries, and guided by the advice of experts in each subject area, Cambridge University Press is using state-of-the-art scanning machines in its own Printing House to capture the content of each book selected for inclusion. The files are processed to give a consistently clear, crisp image, and the books finished to the high quality standard for which the Press is recognised around the world. The latest print-on-demand technology ensures that the books will remain available indefinitely, and that orders for single or multiple copies can quickly be supplied.

The Cambridge Library Collection brings back to life books of enduring scholarly value (including out-of-copyright works originally issued by other publishers) across a wide range of disciplines in the humanities and social sciences and in science and technology.

A Synopsis of Science

From the Standpoint of the Nyaya Philosophy

VOLUME 1

JAMES R. BALLANTYNE

CAMBRIDGE UNIVERSITY PRESS

Cambridge, New York, Melbourne, Madrid, Cape Town,
Singapore, São Paolo, Delhi, Mexico City

Published in the United States of America by Cambridge University Press, New York

www.cambridge.org
Information on this title: www.cambridge.org/9781108056328

© in this compilation Cambridge University Press 2013

This edition first published 1852
This digitally printed version 2013

ISBN 978-1-108-05632-8 Paperback

A
SYNOPSIS OF SCIENCE;

FORM

THE STANDPOINT

OF

THE NYÁYA PHILOSOPHY.

SANSKRIT AND ENGLISH.

VOL. I.

*

PRINTED FOR THE USE OF THE BENARES COLLEGE
BY ORDER OF GOVT. N. W. P.

———

𝕸𝖎𝖗𝖟𝖆𝖕𝖔𝖗𝖊:

ORPHAN PRESS:—R. C. MATHER, SUPT.

1852.

ADVERTISEMENT.

In order that the precise design of this Synopsis of Science may be understood, the compiler has been directed to recapitulate briefly the considerations, detailed in the published Educational reports of the last five years, which have guided his proceedings in regard to the Sanskrit department of the Benares College.

On receiving charge of the Sanskrit College it was my duty to become acquainted with its constitution and history, with a view to introducing whatever improvements might be found consistent with the retention of its character as a seat of Sanskrit learning not unworthy of its classic locality. I was not required to give an opinion whether the funds devoted to the encouragement of Sanskrit learning might be profitably diverted to other purposes. The Sanskrit College was designed to be a permanent institution, and it was not needed that I should deprecate the idea of foregoing in any degree the peculiar advantages offered by such an institution for the advancement of education in India,—advantages for which there could be found no substitute. The proximate end, on which these peculiar advantages seemed capable of being brought effectually to bear, is the development of a language adequate to the reproduction of European thought, and the construction of a scientific literature rightly adapted to our educational purposes, by being in a form congenial to the Hindú mind, and free from barbarisms of speech. It was through the Sanskrit, I perceived clearly, that this must be effected ; but it has never been contemplated that in the Sanskrit the results should remain locked up. Such then being the proximate end, it is here to be shown, in chronological order, how the end has been kept in view, and what pro-

a

gress has been made in sketching a design which would require many hands for its completion.

From a letter of Mr. Jonathan Duncan, the Resident at Benares, dated 24th December 1798, it appears that the Sanskrit College was at that time founded " for the cultivation of the laws, literature, and (as inseparably connected with the two former) religion, of the Hindoos." The discipline of the College, it was added, was " to be conformable in all respects to the *Dharma S'ástra* in the chapter on education." These terms, as I remarked in the Report for 1846-47, " appear to contain the germ of nothing beyond the conciliating of the natives of India by paying a graceful compliment to their language and literature, and of perhaps providing better educated Pandits to act as legal counsellors than could otherwise have been always met with. For many years all the efforts of the various gentlemen who took an interest in the College appear to have been directed to the increasing of its efficiency in these respects." The first decided effort which I found recorded for turning the institution to further account was that of Mr. J. Muir, C. S., who undertook the duties of Principal during the session of 1844. Mr. Muir delivered lectures, in Sanskrit, on Moral and Intellectual Philosophy ; and the sentiments which he then inculcated have often since that time furnished topics for discussion in the College. On the year 1846-47, my impression, as recorded in the printed report, was this, that in the studies of the Sanskrit College all improvement at present must be in the way of addition, not of substitution, because " The most perfect European education bestowed upon a young Bráhmin, however great a blessing it might be to himself, would exert no beneficial influence beyond his own breast, if unaccompanied by the amount of Sanskrit erudition which is indispensable for securing any degree of respectful attention to his words."

The most interesting experiment of the session 1847-48 was the introduction of the study of English into the Sanskrit College. The repugnance of the pupils to this new study was overcome by the offer, to the Senior Students whose period of study was expired, that they should be allowed to retain their scholarship allowances on condition of their reading English. These men I was very anxious to retain. They had reached a point of mental culture at which they had become worthy of being reasoned with on the comparative merits of the civilization of ancient India and of modern Europe,—a point which the junior pupils were not likely to reach until they in turn should be past the age at which they could retain a scholarship under the existing regulations. Government having sanctioned the proposed experiment, the English class of pandits was formed. Its aspect at the opening of the session, as remarked in the printed report, " was not auspicious. The majority of the pupils were very averse to the study, and seemed to think themselves in some measure degraded in the eyes of the other students. They attended reluctantly, when every device for evading attendance failed :—books were lost or had not been supplied ; pens and ink became suddenly unprocurable ; and half the time allotted for the lesson was sometimes spent in settling the fastidiously protracted preliminaries. The pandits seemed greatly to dread being desired to attend in the English College Bungalow, where their slender acquirements in English might expose them to a disadvantageous comparison with little boys who had been reading for a year or two. When they found that no such design upon them was really contemplated, their apprehensions gradually wore off, and ultimately they came over (from the building appropriated to the pandits) to the English department, of their own accord, for several hours daily, in order that they might be within reach of assistance when preparing their lessons." The first want of this class was a suitable English grammar, all the existing grammars being, with reference to the pandits, at once redundant and defective, inasmuch as these manuals take for granted that the learner knows nothing of grammar as a science, and that his vernacular is English or a language of similar idiom. I therefore prepared an outline of English grammar in Sanskrit, which was communicated to the pandits in the shape of lectures, and, after having run the gauntlet of their by no means indulgent criticism, was printed for the use of the class.

In the session 1848-49 steps were taken for bringing about some mutual understanding between the students of the Sanskrit and of the English departments respectively. It was noticed, in the report on the year, as a fact to be lamented, "that the advanced scholars of the English and of the Sanskrit Colleges, though speaking the same vernacular, are mutually unintelligible when the conversation turns on the subject of their studies. The technical terms with which they are respectively familiar, being the product of opposite theories, are not convertible by one who is not conversant with both.

"The consequence is, that the Pundits, in full reliance upon a dogmatic, and, as they think, inspired philosophy, which has stood the discussion (such as it has yet encountered) of centuries, look with calm superiority on the pretensions of a more modest philosophy, which avows that it is only progressing towards that perfection which it cannot hope to reach,—whilst on the other hand our English Students, struck by the imposing methodical completeness of the Brahmanic systems, which they cannot comprehend in detail, and bewildered in every attempt to cope with the dialectical subtlety of the Pundits, who, they see perfectly, though unintelligible to the English Student, are quite intelligible to each other, become possessed by an uneasy feeling, that there is more, if they could but come at it, in the Sanskrit philosophy than is dreamt of in ours. Hence comes the apparent anomaly that a man who can expound the Newtonian Astronomy, consults his astrologer with the same deference as the most ignorant villager ; and confusedly believes in his heart, what the Jesuit Editors of the "Principia" only professed with their lips, that the earth stands still, though the hypothesis of its motion may suffice to account for the phenomena. Hence it is also, that although acquainted with the theory of eclipses, and able to calculate them by European formulæ, he would not on any account neglect to perform the ceremonies ordained for the purpose of helping the luminary out of the jaws of his mythological enemy, the trunkless demon of the ascending node. The only way to remedy this, is to put such a one in a position to judge for himself by making him sufficiently well-acquainted with both sides of the case. It is scarcely necessary to observe that a decision in our favour carries ten-fold moral force with it when it is known that the person so deciding knows not merely what he embraces, but also, thoroughly, what he deliberately abandons."

With the view of enabling the students of the English department to meet half-way the Sanskrit students who had began the study of English, two steps were taken during the session 1848-1849. The one was the preparation of an English version of the Sanskrit school-grammar, the *Laghu Kaumudí*, with references and comments :—the other was the delivery of a set of lectures on the Nyáya Philosophy. As it may strike the reader, if unacquainted with the subject, that the translating of the *Laghu Kaumudí* was a superfluous labour while English grammars of the Sanskrit already existed, I may be permitted to quote the opinion of Professor H. H. Wilson, to whom the first fasciculus (of about a hundred and fifty pages) had been sent. Professor Wilson says of the version, " It will be of infinite use "to those who wish to learn Sanskrit substantially and in earnest. I " went through the text, as far as your translation extends, with one of " my Oxford pupils, to his great gratification and advantage." The peculiar advantage of studying the Sanskrit grammar in the shape in which it is presented in the *Kaumudí* is this, that the learner is thus pre-pared to avail himself of the rich treasures of Sanskrit philology, which, to the mere reader of the grammar reduced to the European form, are a sealed book. That the philological works of the Hindús contain much that is yet to be gained from them may be inferred from the anxiety expressed by Dr. Max Müller, the editor of the Vedas, that the " Great Commentary" on PÁNINI's Aphorisms should be printed. An edition, sanctioned by Government N. W. P., is in progress here.

The text-book taken for the lectures on the Nyáya Philosophy was the *Tarka-Sangraha*. Here, as remarked in the report, "I took sentence by sentence, first giving the original, which my hearers were able partly to understand; then a translation ; and then a commentary, pointing out the correspondence of each part to the several divisions of European sci-ence, and noticing anything analogous in the speculations of antiquity that occurred to me as likely to do good, either by showing that the same truths had been hit upon, or the same errors for a time adhered to, *out* of India as well as *in* it. These lectures were listened to with mark-ed interest, the subject being one which the students are ambitious of understanding,—one which can easily be made clear to them with the aid of explanation in English,—and one which the pandits have not the most distant conception of the possibility of explaining in an exoteric fashion." The Lectures on the *Tarka-sangraha*, including the text and translation, were printed for the use of the College. The discussion which the translation has undergone has suggested various amendments, and an improved translation (accompanied by a version in Hindí) has had the advantage of being superintended, in passing through the press, by my valued coadjutor Mr. F. Edward Hall.

At the annual examination of 1848-49, thinking that some account of the actual nature of an examination of the Sanskrit College might prove interesting, I selected such questions in each department as might give a general idea of the topics which occupy the attention of the native lite-rati. Some of these here follow.

QUESTIONS ON THE VEDANTA OR THEOLOGICAL SYSTEM OF PHILOSOPHY DEDUCED FROM THE VEDAS.

" What answer do you give to the objection that if the Divine Spirit be, " as you say, devoid of qualities, it cannot be made the subject of medita- " tion ?"

" Determine what is the real state· of the case in regard to the following " doubt, viz., whether an elephant and the like, seen in a dream, is or is not " produced at that time, seeing that it has no material cause ?"

" How is the opinion of the *Vaiseshika* School of the *Nyáya* sect respect- " ing atoms, to be refuted ?"

" Explain the erroneousness of such opinions as that of the soul's being in " the form of an atom ?"

QUESTIONS ON THE SÁNKHYA SYSTEM OF PHILOSOPHY (OF WHICH THE ORTHODOXY IS HELD TO BE RATHER QUES-TIONABLE).

" How do you prove that primeval nature has an independent existence ?"

" How do you meet the arguments of the Vedantists who deny a duality in " the universe, asserting that nothing exists except God ?"

" Prove the existence of the Deity according to the views of the Theisti-
" cal school ?"

" How does the soul, in transmigration, enter another body ?"

" By what ascetic practices is a knowledge of the past and the future at-
" tainable ?"

" How do you know that Quietism is right, and in accordance with
" Scripture ?"

QUESTIONS ON THE NYÁYA OR LOGICAL AND ATOMIC SYSTEM OF PHILOSOPHY.

" What is the distinction recognized by the *Naiyáyikas* between the Súpreme
" Spirit and the living soul ?—and what objection is there to the opinion
" that, of the two, the latter only exists ?"

" How do you prove that the mind is in the form of an atom?"

" In the opinion of the *Naiyáyikas,* how many kinds of proof are there, and
" what are they ?"

" How do you prove that gold is identical in substance with light and heat ?"

" Before presenting any other questions, it may be well to make some remarks on these. One of the first reflections likely to be suggested by a perusal of these questions is this, that the pupils are taught by one Pundit to
establish, of course by irresistible arguments, positions which the Pundit in
the next lecture-room teaches him to assail and carry by arguments equally irresistible ; and this reflection naturally suggests two questions,—does not this
keep the Sanskrit College in a state of feud ? and what is the state of the
Student's mind after he has gone through the incongruous curriculum ? The
first question is easily answered. Provided the pupil reads with a given teacher, that teacher seems to have not the slightest objection to his reading with
any of the other teachers. With the view of determining in some measure
the result of the course of discipline on the minds of the Students, I proposed to the Senior Students the following question.

" ' As the three systems of Philosophy which you have studied in the College
" professedly dispute each other's positions, and cannot therefore all be entirely

" in the right, tell me whether you adopt any one of them to the exclusion of
" the others ; or, provided you really have formed any opinion of your own at
" all, whether you adopt, eclectically, something from each.' "

'' The answers were generally to the effect that all the three systems were re-
concilable with Scripture, and that what appeared in any of them to be a de-
viation from the truth, was in reality only an accomodation to the weakness
of the human understanding, which renders it necessary in the first instance
to communicate the truth under the garb of error ; just as a mother, in pointing out the moon to her child, speaks of it as the shining circle at the end of
her finger, which is intelligible to the child, while the mention of its being
distant by thousands of leagues would have hopelessly bewildered him. This
is plausible ; but the habit of viewing the same assertion as true at one moment and false the next, has apparently helped to lead to the existing remarkable indifference as to what is true in itself. Truth, under such circumstances,
becomes a matter of taste, concerning which " non est disputandum," except
in so far as this disputation may furnish matter of amusement or display. We
are not here enquiring into these curious philosophical systems as a mere matter of curiosity. The question of questions in regard to them is here—how,
and how far, they are capable of being turned to account.

" Of the three leading Schools, the *Veddnta,* the *Sankhya,* and the *Nydya,*
the first, being an attempt to reconcile Hindú Scripture with Philosophy, ob-
viously does not promise much to aid us. The second is as nearly as possible
a system of Nihilism, though its advocates protest against imputing that character to it. It contains much that is ingenious, and not a little (as Professor
Wilson and others have shown) that has been only recently excogitated in
Europe. But as a system, it tends to nothing that we can have any interest
in promoting. We cannot make its plan therefore the ground-work of any
curriculum of our own. The *Nydya* on the other hand is a very fair, and, in
some respects, admirable, attempt, on the part of certain speculative philosophers, who had made perhaps as many observations and experiments as they
had opportunities of making, to present a complete and consistent physical as
well as metaphysical theory of the universe. Of this system, therefore, I have
chiefly made use, in laying the foundations of an attempt to present to the
Students of the Sanskrit College an equally comprehensive view of the universe, divested of those errors in their own *Nydya* which modern observation
and experiment have shown to be such, and giving somewhat of its due promi-

nence to the physical departments of science, which were much less prominent in the original exposition of the *Nyaya* doctrine than its metaphysics, to which the physics were entirely subordinated as they have ever since remained. While their system professes to embrace the universe, it really neglects all that forms the subject-matter of the physical sciences, and consequently its professors look down with self-complacent superiority upon the cultivators of physical science, and with indifference upon its results. The case of Astronomy presents only an apparent exception to this rule, for it is for astrological purposes alone that the bulk of the Brahmins value Astronomy. Here, as in other departments, the knowledge that they have, furnishes too often the main obstacle to their acquiring more. But this is only an additional reason why we should take care to ascertain what it is they have ; for whatever they possess of truth, will remain an obstacle, until we make it an ally.

" The Hindoo mind for a long period, has been in what Whewell calls the ' commentatorial stage,' a stage in which originality is forbidden and shunned. This would seem to present one of the occasions when a just appreciation of the history of an analogous period may be fairly expected to throw light upon the prospect of the future, on its undesirable probabilities, and its more desirable possibilities, possible only if they be properly anticipated. To this consideration I shall have occasion to revert. In continuation of the purely Sanskrit portion of the examination, I subjoin some of the questions on Grammar, Rhetoric, Law, Mathematics and Astronomy. An inspection of these will show that there is here occasion not so much for the Baconian instruments intended to ' originate motion,' as for those that " direct" it when once originated,—the centripetal force, or wooden yoke, of dogmatic authority, having long since converted, what at the outset was onward progress, into the narrow yet interminable orbit of an ox in an oilmill.

Questions on Grammar.

" What is the province of Grammar ?"

" Has a word any sense of its own, or is it merely a mark for the thing sig-" nified ?"

" According to the opinion of the Grammarians (who are at variance on cer-" tain points with the followers of the Nyáya, &c.,) what is the real state of " the case in regard to this sentence, viz., Yajnadatta cooks rice ?"

" Many other questions were given on this branch, but they related chiefly to the technical treatises on the etymological structure of the language, one of which treatises* I have already mentioned my wish to render accessible to the English reader.

QUESTIONS ON RHETORICAL COMPOSITION.

" What are generally the faults that can be committed in the way of com-
" position ?"

" What faults can be committed in regard to the management of a simile ?"

" What is the difference between a simile and a metaphor ?"

" Give some account of the cases in which a metaphor is of such a nature
" that the thing for which it stands need not be further hinted."

QUESTIONS ON LAW.

" Two full brothers are co-parceners, one of them dies childless, leaving a
" wife ; afterwards the other who has lost his wife also dies, leaving a daughter
" who is childless, but whose husband is living. State what right to the pro-
" perty belongs respectively to the widow and the daughter, according to the
" *Mitákshara*, and also according to the *Dáyabhága*."

" A person seated in his carriage drawn by a horse, and driven by an expert
" driver, is proceeding along the road, and shouting ' keep out of the way,
" keep out of the way !' As luck will have it, a man is driven over, and kil-
" led : which is to pay the penalty, the driver or his master ?"

" A she buffalo, with her calf, being entrusted to a keeper during the day
" time, having eaten another person's corn, lies down to sleep there unhinder-
" ed. In this case what penalty must the owner of the buffalo pay ? How
" much is the owner of the field entitled to receive ? Is the keeper blameable
" or not, and if blameable what penalty must he pay ?
" What kind of proof is most effective, and in what kind of cases ?"

" Being curious to see what the Students would make of a case for which
they could find no precedent in their law-books, I proposed the case which
Reid cites as an example of an insoluble dilemma,—of the sophist Protagoras

---

* The *Laghu Kaumudi*—since printed.

and his scholar. Just as I expected, they tried it by every one of their technical rules in succession, never doubting, but that one or other of the keys must fit. When they found, to their great surprise, that this was not the case, they betook themselves to the unusual task of unaided thought; and whilst one decided that the judge must decree in favor of the pupil, another said that he must decree in favor of the master, and a third that he had better dismiss the case without giving any opinion on the matter, which last is the resolution that the Greek judges are related to have come to. The law Pundit, to whom these opinions were submitted, took two days to consider the case, which he also tried in vain by his body of rules which never had failed him before. At this he made no secret of his admiration, but at last he hit upon a solution not uncreditable in my opinion to his sagacity, viz., that the pupil was decidedly entitled to a verdict in his favor, and that then this would furnish good ground for a new action in which the teacher must needs gain his point. I mention this as illustrating (what I wish I could illustrate by instances of a character less slight) the lively and salutary excitement which may be created among the Pundits when any thing that they really take an interest in, is presented to them in such a way as to compel them to step out of the beaten track. Unfortunately, in regard to those subjects respecting which their knowledge is most defective, the difficulty is to get them to take any real interest at all. The method which I have found to answer best, is to take as a starting point some established point in their own philosophy, and to show how the philosophers of Europe have followed up the enquiry.

" For example, I found that the Pundits entertained a very low opinion of the European Logic, some account of which had been supplied to them from the popular work of Abercrombie. On this subject I perceived that all my explanations were thrown away, until it occurred to me to enquire carefully whether the knowledge of my hearers did not stop short at some point between which and the knowledge that I wished to communicate, there remained some gap to be filled up, before they could discern that the one was but the continuation of the other. The result was extremely satisfactory. The Pundits, gratified by the admission that their own view of the process of inference is correct so far it goes, laid aside their jealous hostility, which was succeeded by lively curiosity to know how the thing could be carried further;—and thus was obtained, what was wanted, an unprejudiced hearing for what was to be brought forward. It is worth noticing that the very apparatus of technical rules—the " *Barbara Celarent,*" &c.,—which now repels so many in Europe,

was hailed at once as an earnest of there being something valuable in the treatise shown to them. The contrivance of significant vowels and indicatory consonants was at once recognised as akin to that of *Pánini* in his institutes of
Sanskrit Grammar, and the fact that the system had been matured more than
two thousand years ago, invested it with another charm in their eyes.

" These things appear to be worth bearing in mind, for they would seem to
ndicate that the likeliest way to get the Pundits to lend an unhostile ear to
what we have got to say, is to lead them from the very point to which their
correct knowledge has attained, as much as possible, by the steps which the
European mind itself took in reaching its present conclusions after starting
from an analogous point. For example, having secured the attention of a set
of Pundits to the Aristotelian Logic, and having thereby gained something of
additional respect in their eyes, I explained to them the design and character
of the *Novum Organon*, and pointed out which division of their own philosophy,—a division avowedly the least satisfactory of all as hitherto treated
by their own authors,—is represented by this great work. I have found no
work the general description of which has more excited the curiosity of the
most intelligent of my Pundit auditors than this. Of the way in which we
are making use of it, I shall have to speak when narrating the studies of the
Anglo-Sanskrit class. Bacon himself, though as a classic he will always be
read, yet is out of date in Europe as the actual guide in scientific investigation. The employment of his own instrument has enabled subsequent enquirers to detect his own deviations from the right track of discovery : but this
very fact, if it be carefully kept in view and properly made use of, gives additional value to his writings as an instrument for promoting the intellectual
advancement of India."

During this session, of 1848-49, I delivered to my class of pandits
part of a course of lectures, in Sanskrit, on ' The Mutual Relations of the
Sciences.'* At the examination the following were—

---

* The topics touched upon in these Lectures were—Part 1—Astronomy, Geography, Zoology, Botany, Mineralogy, Geology, Chemistry ;—Part 2—Arithmetic, Algebra, Geometry, the Calculus, Mechanics, Hydrostatics, Pneumatics,
Acoustics, Heat, Optics ;—Part 3—Metaphysics and Mental Philosophy, For-
mal Logic ;—Part 4—The Philosophy of Investigation, Grammar, Rhetoric,
Ethics, Law, and History. The four Parts are printed in Sanskrit and English.

QUESTIONS ON THE LECTURES OF THE SESSION.

" How many planets are there? what is the form of their " orbits ? and around what do they revolve ?"

" What is the form of the earth ? what proportion of its surface is occu- " pied by the ocean ?"

" Marine productions are sometimes found in mountain ranges :—account " for this."

" Of what description was the fossil elephant ?"

" What is the chemical composition of water, and of atmospheric air ?"

The class to whom these questions, in writing, were put and answered in Sanskrit, Mr. D. F. M'Leod, C. S., examined orally on the English books that had been read during the session. Mr. M'Leod remarked as follows :—

" I was present at, and a party to, the examination of this Class in the " ' Moral Class book,' Bacon's 'Novum Organon,' and other subjects, inter- " spersed with questions on Grammar, and was very highly gratified by the " result. The acuteness and profundity acquired by these Scholars in the " course of their Sanskrit studies, is carried by them into their English ones, " and brought to bear, with great effect, upon every branch of knowledge in- " troduced to them through that medium. Several of them read with consi- " derable fluency and precision, and though, from the comparative briefness of " the period which has elapsed since they first commenced it, and the great " difficulty of English orthography and pronunciation, much cannot be expec- " ted from them in this respect,—yet, from the answers given, and the mode " of treating the subjects adopted by them, the impression is irresistibly forced " on the examiner that the knowledge they have acquired is in reality greater " than it at first appears : the converse probably, of what might with some " justice be said of most ordinary Classes."

" The Class is, in my opinion, the most interesting and important in the " whole institution. If carried on as it has been commenced, it affords every " promise of realizing the expectation entertained of it by the Principal, with " whom exclusively it has originated ; and I most sincerely trust that its aim " and object may never be lost sight of, until the experiment shall have had

" the most complete fulfilment, and its results been exhibited in an unmistake-
" able form."

In the course, above referred to, of lectures to this class of pandits, I followed the division of the sciences adopted by Dr. Ar-
nold in his address to the Rugby Mechanics' Institute.

" But when we arrived at the questions of Metaphysics, the consideration of
which I had sought to postpone until we should have gone amicably over some
less debatable ground, and thus perhaps have obtained some previously established matter to serve for reference to when illustrations were required, my
auditors, accustomed to that strictness of methodical arrangement which is so
attractive in their own systems, immediately began to object, (like Demetrius,
in the play, to Moonshine with his dog and his thorn-bush—) " Why, all
these should be put in the lantern :"—in other words, that what in Europe is
treated as a branch of science, under the name of Metaphysics, ought, with
some fitter name, to furnish, like their own *Nyáya*, the framework of the
whole of the sciences. Having foreseen this objection, I stated to my critics
my reasons for having adopted a different order in addressing them, and mentioned my hope that there might be furnished, for the satisfaction of those
who felt interested in the preliminary course, a fuller exposition of the sciences, with an arragement modelled on their own.

The Synopsis to which these recapitulatory notices are introductory is
an attempt to redeem the pledge thus given. Its principle of construction
will be the better understood from the perusal of an extract from the Minute of Mr. M'Leod on the examination of 1848-49. In that Minute,
Mr. M'Leod recommends attention to the Report for the year,—

" More especially as, in conjunction with those which have immediately preceded it, it developes the principles and gradual progress (in its application)
of what I believe may be considered as an almost entirely novel theory of education, as applied to India, or any other nation similarly circumstanced in
respect to its instructors ; or at all events one, which in one important respect
differs greatly from that which prevails in our other educational Institutions
generally.

" Those who have heretofore had the direction of Educational measures in this
country—whether on the part of Individuals, Associations, or the Government,

appear to have acted for the most part on the principle of regarding the Hindoo mind, for all practical purposes, as a "tabula rasa" in respect to any preconceived ideas, and pre-established system of literature, philosophy, or science either useful and valuable in themselves, or esteemed such by the people with whom we have to deal: and the effects of this appear to me to have been highly prejudicial in many ways; as I think a survey of the general results at our presidencies, as well as elsewhere, will satisfy most candid observers.

"It has tended to segregate from the mass of their countrymen the elêves of our Schools and Colleges; and these, finding that they have no longer ideas in common with those of their brethren who have not been similarly educated, but are rather contemned by a large portion of them, at the same time that they are conscious of being more favorably regarded by the members of the ruling nation, and more nearly assimilating to them in sentiments, have very generally evinced a disposition to regard the former with contempt, and to imitate the least commendable of the peculiarities of the latter; a self-sufficient assumption of superiority taking the place of the humility which a mere entrance within the portals of the vast field of knowledge might be expected to produce. It has also greatly incapacitated these youths for the task of communicating to their countrymen the knowledge which they have themselves acquired, even if other circumstances favored the endeavour; so that except to whatever extent circumstances may in any locality have given extension to the direct study of English, little or no progress has as yet been made towards inoculating the mass with the knowledge of the west; and lastly it has entirely repelled from us, by wounding their self-esteem and pride of learning, those classes who possess, and who unless their position be more stratagetically stormed, I doubt not will yet long continue to possess almost unbounded influence over the large majority of the nation."

Mr. M'Leod goes on to express his hope that, by carrying out the system laid down in the report, the unsatisfactory state of things then existing may be reversed; that "the student of the European school may be brought to understand, appreciate, and sympathize with the Oriental scholar, and the latter with the former;" that "the analogous or identical truths of the systems respectively pursued by each may be traced out and established as common starting points;" and that thus the learned Hindús "may be conciliated and gradually won over to our cause, and their great erudition and philosophical training brought to bear with effect and power upon the researches which we most value;" These sentiments of Mr. M'Leod, most accurately embodying the drift of the Report, will suffice to explain the general aim of the following Synopsis. Why the aim has been taken exactly as it has been, it remains to explain.

In the Report for 1849-50, speaking of the English class of pandits, I remarked that—" Aided by these men I shall now listen with simple disregard to the discouraging reiterations of those who insist that the truths of science and of philosophy cannot be communicated to the Hindús without the use of words which would go to barbarize their language ; as if a language richer in roots than any European one, and far more finely organized, could not supply as many available terms ;—as if the Sanskrit needed instruction at the hands of its grandchild the Greek. To render intelligible our plan of operations for the next session, I may here remark that my first attempt to open a communication with the frequenters of the Sanskrit College was made in the shape of a set of lectures on the circle of the Sciences. The Sanskrit version of these was carefully revised by Pandit Bápú Deva, whose rendering of many of the scientific terms was most felicitous. I learn that these renderings have been incorporated into the English and Sanskrit Dictionary now preparing by Professor Williams for the use of the College at Haileybury. In those portions of the lectures which related to sciences which the pandit had not studied, we were less succesful than in the others. To ensure success it would have been indispensable to investigate the first sources of the nomenclature appropriated to the same or kindred topics in the Hindú Philosophy ; and how I propose that this shall be done, I now proceed to state.

" The multitudinous treatises in the six great schools of Hindú Philosophy are all based upon, and are held to be of authority only in so far as they coincide with, the six collections of Aphorisms promulgated severally by the founders of the six schools. The aphorisms were intended not to convey the doctrine but to record it ;—hence their oracular brevity. They resemble, in some measure, such formulæ as that of the Binomial Theorem, which, when once explained, enables one to recollect readily a complicated series of facts which—if expressed in words at length—no human memory could have retained a knowledge of." The Aphorisms, therefore, from the first, had been attended by a comment, oral or written ; and, with the aid of these, it was now proposed to make a critical examination of the original Aphorisms, because, "only by tracing the development of Hindú thought, and of the terminology in which it clothed itself, can we hope to avoid completely all such misappropriation of terms as that which has, to a certain extent, baffled all European attempts at translation into the Hindú dialects wherever the subject of discussion transcended the palpable."

It has been already remarked, that, of the Hindú systems, the Nyáya is the one which lends most readily its arrangement and terminology for the conveying of new scientific information intelligibly and satisfactorily to a learned Hindú. Whilst, therefore, the translation of the several systems has been carried on simultaneously,—each system supplying terms and views which will be turned to account,—it was the arrangement of GAUTAMA's Aphorisms that was adopted as the frame-work of the proposed Synopsis. In his 1st Book (—which, along with the 1st Book of the *Vaiseshika*, the *Mímánsá*, and the *Vedánta* Aphorisms, has been printed with a translation for the use of the College—) GAUTAMA lays down the plan of the whole Nyáya system. This he effects in sixty Aphorisms. He starts with the grand question of questions—the enquiry as to how we shall attain the *summum bonum*—the "chief end of man." This he declares, in his 1st Aphorism, can be reached only through knowledge of the truth. But have we *instruments* adapted to the acquisition of a knowledge of the truth? We have our senses &c. ; and these GAUTAMA enumerates and describes in Aph. 3—8. But, if we have instruments, let us know what are the *objects* to which these are appropriate. GAUTAMA replies to this in Aph. 9—22. But the bare enumeration and definition of objects does not ensure a correct and believing knowledge of them. The state intermediate between hearing and believing—viz. doubt—he defines in Aph. 23. But how is a man to get out of Doubt? He will be content to remain in doubt if there be no *motive* for enquiring further. Here—Aph. 24.—he takes occasion to explain what constitutes a Motive. But in every enquiry, to reach the unknown,

c

we must start from the *known ;*—there must be *data.* These be describes and classifies in Aph. 25—31. The data being determined, it is
proper to determine the order of procedure in demonstrating thereby
something not granted. This he sets forth in Aph. 32—38. But, thus
far, we have been shown an arrangement for hearing only one side
of a question, and how can we be sure that the opposite side is not the
right one? The propriety of hearing both sides of a question
before making up our minds GAUTAMA suggests in Aph. 39-
40. But an honest enquirer may have heard both sides and still be in
perplexity. Is he to be turned adrift? Not at all. Candid discussion
with one who holds the same first principles is open to him ;—Aph. 41.
There are yet others, besides honest enquirers, that are not utterly to be
rejected. A person, not hopelessly irreclaimable, may *wrangle* for the
sake of a seeming victory. In Aph. 42, therefore, he defines wrangling.
A person, still perhaps not hopelessly irreclaimable, may descend lower
than the former by carping at others without undertaking to settle anything himself. In Aph. 43, therefore, he defines cavilling. Wranglers
and Cavillers, in default of good reasons, must make use of *fallacies.*
The various forms of Fallacy, therefore, he defines in Aph. 44—49. But
whilst there are fallacies by which a man may deceive himself as well
as others, there are *frauds* which are employed only dishonestly for the de-
ception of others. These he describes in Aph. 50—57. Descending a
stage lower, an opponent may employ objections so futile as to be capable
of deceiving no one. It is well to know in what consists the futility of
such objections. This he shows in Aph. 59. Finally an opponent, sink-
ing even below the former one, (who *knew* what he was opposing, though
he could make none but a futile opposition), may be unable to understand the proposition ;—Aph. 59-60. Here GAUTAMA's patience is
exhausted, *but not before.* Against everything but the invincible combination of the spirit of contradiction with *stupidity,* he seeks to arm
himself at all points. An objection the most frivolous—or even futile—
provided it be tendered by one who understands the proposition—he does
not refuse to deal with. The objection might perplex some honest enquirer, and therefore GAUTAMA, or the follower who has imbibed his spirit,
does not consider himself at liberty to consult his own ease by scouting it,

though he himself may see its futility plainly enough. It is fair to re-
member this when we meet with ludicrously frivolous objections gravely
treated in a Nyáya work. The author is not to be supposed to have *in-
vented* the objection. It was offered to him—offered very possibly for
the purpose of vexatiously puzzling and perplexing,—and the Naiyáyika
will not allow himself to be puzzled and perplexed. The most cavilling
opponent is not to be allowed the semblance of a victory; he shall not
be allowed to boast even of having put the philosopher out of temper.
This single triumph—such as it is—is reserved for the absolute blockhead.

Now, I should be glad to learn from those who undervalue the *method* of
the Nyáya—speaking of its exposition as " tedious, loose, and unmethodi-
cal"*—how could that method be much improved? One must not
imagine that he has answered this question when he has shown that there
are some important matters not here explicitly noticed by GAUTAMA.
He must be able to show either that there are important matters for
which the system provides no place, or that the order of procedure is mis-
arranged. The order of procedure, according to my own view of it,
I have explained. The enquiry whether there is anything within the
range of conception, for which the arrangement above sketched does not
furnish its appropriate place, is one to which we aim at giving a practical
reply in the Synopsis now commenced. Meantime let us take a cursory
glance at the range of topics which it will be necessary to deal with in
preparing a consistent digest of European knowledge for the use of India.

Knowledge may be subdivided, at the outset, into that which is communicated by Revelation, and that which is searched out by Philosophy. It
is with the latter branch that we must at present be concerned. The knowledge that is searched out by Philosophy constitutes Science; and, when applied to the attainment of further ends, it gives rise to Art. Science relates
to things that exist as realities, or to their modes of existence. What exists
as a reality is either Matter or Spirit. The modes of existence, as Number,
Magnitude, and Motion, give rise to such sciences as Arithmetic, Geome-

---

* Ritter's History of Philosophy—English version—vol. iv. p. 366

try, and Mechanics. Matter may be regarded as Imponderable or Ponderable. Imponderable matter presents itself in the shape of Heat, Electricity, and the like. Ponderable matter may be regarded either as in Atom or in Mass. In Atom, it furnishes the object-matter of Chemistry. Viewing it in Mass, we have the earth either considered in connection with the heavens, by Astronomy ; or considered alone ; superficially, by Geography ; or as to the causes of the existing distribution of its parts, by Geology ; or in regard to its constituent masses. These are either Inorganic, in which case they belong to Mineralogy ; or Organic, the latter division including the Vegetable and the Animal kingdoms. Vegetables are considered structurally in Botany, and functionally in Vegetable Physiology. Animals are considered structurally in Zoology and Anatomy, and functionally under the two physiological conditions of health and disease,—conditions with which the science of Medicine is conversant.

Reverting to the other main branch of the division,—Spirit may be divided into the Supreme Soul, or God, and the Human Soul. Considered as a philosophical question, apart from revelation, the question of the existence and the attributes of God furnishes the topic of Natural Theology. With regard to the human soul, there arise four questions. We may enquire into its nature and operations, as questions of Psychology ; an enquiry with which it is customary to associate the enquiry into Being—or the science of Metaphysics. Secondly, we may enquire into the duties of the soul, or of its possessor. This gives rise to Ethics, private and public. Thirdly, we may make separate enquiry respecting the instruments of the soul—Language, Inference, and Exposition. Language belongs to Grammar ; Inference to Logic ; and Exposition to Rhetoric. Inference again is either of generals from particulars, or of particulars from generals,—so that Logic is either Inductive or Deductive. Fourthly, Soul, or its possessor, may be regarded as the agent in the events of History ; and this may be regarded either as matter of fact, or as something calling for an exertion of the critical faculty. Viewed under the former aspect, it may be subdivided into Political, Religious, and Philosophical. Considering it critically, we have two questions to ask,—firstly, what are its credentials ; and lastly, what are its lessons.

Assuming that this enumeration includes the topics of a complete liberal education, I am desirous that the whole digest, of which the Synopsis seeks to indicate the starting-points, shall be prepared, in the first instance, with reference, as close as may be, to one of the systems of the universe already current among, and accepted by, the Hindús. In explanation of this, I would beg the readers attention to the two facts, that a mind can be taught only by means of the knowledge that is already in it ; and that a piece of knowledge in any mind —more especially in a mind unfavourably prepossessed—is an obstacle to the reception of any system which, by neglecting to recognize, appears to deny, the truth of that piece of knowledge.* Whatever in the Hindú systems is a portion of the adamantine truth itself, will only serve to baffle our efforts, if, in ignorant impatience, we attempt to sweep it away along with the rubbish that has encrusted it. What kind of engineer should we think him who, when seeking to raise a beacon on the Goodwin sands, should hesitate to acknowledge as a god-send any portion of solid rock, amid the shifting shoals, to which he might rivet one of the stays of his edifice ? When a headstrong opponent of an imperfect system treats with indiscriminate scorn what is true in it and what is false, he has no right to complain that his arguments against the false are as lightly esteemed as his scorn of the true. When the Hindús have only halted at a stage short of that which we ourselves have reached, we should rejoice in being able to present to them our superior knowledge as the legitimate development of what is true in their views, and not in the shape of a contradiction to anything that is erroneous.

---

* The scheme of treating the Hindú mind as a tabula rasa, ignoring the existence of Hindú opinions right or wrong, and of attempting to educate India solely by means of the English language, I of course regard as unfeasible. The hopelessness of this somewhat indolent and much too supercilious scheme has been shown by Mr. B. H. Hodgson, in his letters on Indian education, with a clearness that might convince prejudice itself. Let there be as much English education given as possible,—the more the better ;—but let not the delusion be cherished that we shall then have done our part.

It is with such aims that I have made use of the Nyáya system as the framework of the following Synopsis. Wherever in the 1st Book, I have been constrained to dissent from GAUTAMA, I have recorded the amount of divergence in a note. The English reader will perhaps be surprised to find in how few instances I have considered it necessary to record dissent. He may probably think that I ought to have done it in more; but, in regard to this, I trust I shall be able to satisfy the attentive reader as the work proceeds. The style of the commentary will strike the English reader as stiff and ungainly. If he will compare it with the translation of GAUTAMA's Aphorisms which is interspersed with native comment, he will see that I have intentionally imitated the established style of exposition. My object was, not to introduce a new style, but to convey truth in the style which, as being the established style, was the least likely to provoke cavil.

After completing the first Book, containing the plan of the whole system, I did not consider myself bound to follow the order of GAUTAMA in the development of the matters therein propounded. Having come to an agreement with my Hindú readers as to how much of the first Book we jointly accepted, the development of the accepted portions might of course, without detriment to the mutual understanding between the parties, be carried out in such order of succession as circumstances should render advisable. Now it happens that the Nyáya Philosophy, though furnishing a framework for an encyclopedic body of doctrine, has, in practice, confined itself very much to the topics of Logic and Metaphysics, to the neglect of the topics of Physical Science. This defect it appeared desirable to remedy in the first instance; and, accordingly, my 2nd Book, after a concise account of the Five Senses (in regard to which some erroneous opinions of the Hindús are there corrected), proceeds to treat of "Matter, or that—the qualities of which furnish the objects of the Senses." This leads to Astronomy, Geography, &c. These Sciences are here treated very briefly, because, as I intend the Synopsis to be employed by the Nyáya Professor, and to furnish for his pupils a comprehensive view of the relations of all the principal sciences to the philosophical system of the Nyáya, the main design would be obscured if each topic were treated much more fully. For the use of the other classes respectively

I intend that each of the sections in the Synopsis on Geography, Chemistry, Mechanics, &c., shall form the starting-points of more expanded treatises. The completion of these would require such resources as those of the College of Translators which Mr. B. H. Hodgson has designed, and which I shall not despair of seeing established. In urging the advisableness of preparing a carefully systematized body of instruction, it is of course implied that the irregular and immethodical plan hitherto followed for supplying the desideratum does not appear to me calculated to produce effects correspondent to the labour thus immethodically expended. The plan hitherto has been to say, "Here is a good book—let us translate it." And so a considerable number of books has been translated, and still the want is far from being supplied,—just because no moderate, much less any undigested, selection of books can give the quintessence of the Library of Europe. One book intrudes largely into the province of another ; and when this is not attended to, we go on reproducing the same thing over and over again, sometimes in a better shape and sometimes in a worse ; and the consequence is, that, while our toil is multiplied, our readers do not add to their knowledge in proportion as they proceed with their reading. The evil of such a state of things in Europe is attempted to be remedied from time to time by the issue of Encyclopedias, which undertake to sum up the total of real knowledge up to the date of publication. The adoption of such a comprehensive method at the outset would have saved, in Indian education, a world of time and expense. It is not too late to adopt it in regard to the Sanskrit, in which hitherto next to nothing has been done to make those conversant with that language acquainted with European Science.

To a reader who is not aware of the relation—the still *existing* relation of the Hindú languages to their living and fostering parent the Sanskrit— the only parent to which they can look up for wholesome nourishment*,—

---

* In regard to this point I may transcribe the following observations from an article on ' The Prospects of India, Religious and Intellectual' (Benares Magazine, March 1849,) various portions of which, as conveying opinions

it may seem paradoxical when I assert that the fit preparation of a version of any scientific treatise in Hindí, Bengalí, Mahratta, Guzerátí,
Támil, Telugú, and Sanskrit, is easier than the preparation of the same
set of versions without the Sanskrit one. A little reflection will show
that there is nothing strange in this. What is the difficulty—the tran-
scendent difficulty—in translating a European work into an Oriental language? It is the difficulty of determining the exact amount of corres-
pondency between the different portions of the knowledge, on any subject, elaborated by the East and West, and embodied in their respective
forms of speech. Different philosophic or scientific theories give rise to
different forms of expression; and, where this is disregarded or forgot-
ten, we have the story of Babel repeated indefinitely. Now, this im-
mense difference of cast, both in thought and expression, meets us in
every Indian language which we try to make the vehicle of our knowledge; but if the work which it is wished to communicate to all India is
once put fitly into Sanskrit, the task is well-nigh done. There is little
more difficulty in turning the Sanskrit work into each and all of the

---

which I have seen no reason to change, have been incorporated in the present
Advertisement.

"It is a great and mischievous mistake to regard the Sanskrit in India as a
dead language, in the sense which that term generally suggests. What is
meant by a dead language? Nine men out of ten will reply at once that it is
a language no longer spoken by the people as their mother-tongue. This definition at once suggests to six men at least out of ten, the idea of a language,
the cultivation of which, if desirable at all, is so mainly as a matter of intelligent curiosity. But a very little reflection will suffice to convince any one
that of the languages which, in terms of the foregoing definition, may be
called "dead," by no means the whole fall under the description here suggested. For example, the Anglo-Saxon is no longer spoken by the people of England, and neither is the Latin. To call them both dead languages, however,
does not fairly imply that their claims to attention are equal. In respect of
their influence upon the spoken language of the day, the Anglo-Saxon, from
which, either as a language or a literature, we have long since ceased to gain
anything new, may be regarded as the deceased parent of the English; whilst

vernaculars, when there are properly instructed pandits of all nations at hand, than in turning so many ingots of gold into guineas, sovereigns, and half sovereigns, when the mint is at your command. When a book has been first rendered into unexceptionable Sanskrit, the risk of error, under proper supervision, is at an end ; whereas, if translations are made into each language directly from the English, the risk of misconception perpetually recurs. A correct Sanskrit version is like the golden or platina rod deposited in the exchequer office, by which all the brass and wooden yard-measures in the country can be verified, or rectified. To obviate misconception, it may be proper to add, that, I wish the Sanskrit version to be regarded as the measure and criterion of the *sense*, not as the rigid exemplar of the *form* to be adhered to in the vernacular versions to which it shall supply the matter and the scientific terminology.

As it is in the Nyáya class that the Synopsis is to be employed, there is one section of it which I contemplate treating more fully, in the first instance, than the other sections,—the section, namely, of Logic, Ratiocinative and Inductive. In the Sanskrit Lectures on the Mutual Relations of the Sciences, I furnished the Nyáya pandit with a concise ex-

---

the Latin, from which our language receives yearly accretions, and by whose literature the minds of each successive generation are moulded, acts the part of a living nurse—though we may choose to hold it technically dead. But if the difference be great between these two, which is hidden under the general name of "dead language," much more momentous is the difference which can slip fallaciously out of sight when the same conveniently loose cloak of a generalization based on the non-essential throws its misty folds around the Sanskrit also. The Sanskrit, to all intents and purposes of any consequence, is no more dead than our reader, who would be able to insure his life on his own terms, if he could show that he had the slightest chance of surviving it.

In the Dedication prefixed to the first Edition of his Dictionary, but omitted in the second, Professor H. H. Wilson remarks :—

" The value of the *Sanskrit* language, as an object of literary curiosity, is of no moment, compared with its importance in the light in which it must be

d

position of the European theory of the Syllogism ; and his pupils, while studying it, were not a little struck with the technical apparatus of " Barbara, Celarent, &c.," resembling so closely as it does, in its employment of indicatory vowels and consonants, the time-honoured technical apparatus of the Sanskrit Grammar. In a sketch of " The Philosophy of Investigation," I adopted, from Sir John Herschel's Discourse, the fine example of Induction furnished by Dr. Wells's theory of the formation of Dew. In constructing both of these tracts in Sanskrit, I found it very difficult to satisfy the pandits. The fact being that they had a system of their own, unless all the terms were employed in strict accordance with that system, continual misconception was inevitable. Our modern conception of Induction being that to which is particularly to be attributed our superior progress in science, it appeared highly important, since there existed such a difficulty in coming to a mutual understanding, that the Hindú speculations on the subject should be carefully investigated ; and I was therefore glad at this time to receive from Professor H. H. Wilson a letter suggesting that a translation of the *Anumána Khanda*, or " Section on Inference," the standard treatise on Hindú Logic, would be very acceptable to the logicians of Europe. An examination of the work in question showed me a greater resemblance than I could have hoped to find between the turn of thought and expression in the writer and in Mr. John Stuart Mill, whose work, on " Logic, Ratiocinative and Inductive," I had begun to employ, with good prospect of advantage, in reading with my class of pandits. Two treatises have therefore been commenced, and put to press ;—the one to consist of such a commented abstract of Mr. Mill's work as may best suit at present the class of readers for whom it is intended,—and the other, of a translation of the parallel Sanskrit treatise abovementioned, on the

---

viewed by all who duly consider its connexion with the welfare of this country."

This is an important testimony from a scholar, whose tastes and acquirements would tend to anything rather than to make him underrate the value of a thing " as an object of literary curiosity compared with its value in any other point of view."

same subject, to be studied in the English department also of the College.

For the use of the students in the English department, I prepared, as a sequel to the Lectures on the Nyáya Philosophy, a Lecture on the Sánkhya Philosophy, embracing the text, with translation, of the *Tat-twa-samása*; and a Lecture on the Vedánta Philosophy, embracing the text, with translation, of the *Vedánta-sára*.* These works, having been since printed, are available as portions of a course of Sanskrit study which, it is hoped, may be gradually introduced in the English department as a discipline analogous, in its benefits, to the study of the classics in Europe. A fasciculus of the excellent text-book of the Nyáya employed in Bengal has also been printed, with a translation, for the use of the English pupils; and, to smooth the entrance to the grammar, I have prepared, with the aid of Mr. Hall, a series of "First Lessons in Sanskrit" on the method of Ollendorff, each sheet as it was printed being supplied to the classes. By such means we aim at rendering Sanskrit, with all its fine capabilities as a disciplinal study, no longer a thing here attainable only at the expense of a life-time; and it is thus that we aim further at making the English and the Sanskrit departments of the College understand each other on subjects in regard to which hitherto the students of the two departments, though speaking the same Vernacular, could as little understand each other as the inhabitants of separate planets with separate natural laws. The reader unacquainted with the facts

---

* In the preface to this translation it is remarked as follows :—" That this exposition of the most abstruse of the Hindú systems should be faultless, is very far from probable. In bringing it—as it now stands—before the senior English students, and the class of pandits, in the Benares College, who have for some time been studying English, the intention includes its being submitted to a searching criticism, so that any errors entertained in regard to the system, by Europeans, may have the better chance of being discovered and rectified." The same remark applies to all my other translations. Though no pains has been spared on them, and they are the result of labour curiously disproportioned to the smallness of their bulk, yet I do not presume to offer them to the scholars of Europe except as *proof-sheets* awaiting correction.

may find it difficult to conceive this :—I give him my word for it. Another standard work, the treatise on literary criticism entitled the *Sáhttya Darpaṇa*, has been selected as a portion of the course of Sanskrit study designed for the English department. The printing of this work, with its translation, as the subject seemed likely to interest European scholars, has been undertaken by the Asiatic Society; and the first fasciculus is now published as a portion of the 'Bibliotheca Indica.' I trust that in due time the whole of the course of study pursued in the Sanskrit department will become by such means opened to the pupils of the English department; and I am not without hopes that when this shall be the case, a collateral result will be the removal (from the minds of those competent to judge) of a considerable amount of misconception at present existing in regard to the nature and value of a Sanskrit education. The Oxonian who rightly esteems Aristotle will not disesteem Gautama and Śiromaṇi when he becomes equally well acquainted with them.

In the Benares College,—besides the Arabic and Persian department, which does not come within the scope of the present notice,—there is another department not yet adverted to, viz., the Hindí. According to the native system the young Bráhman is confined to the study of the Sanskrit grammar for a protracted period before he learns anything besides the sounds and inflections of words. This arrangement being objectionable, the younger students of Sanskrit are now formed, during certain hours of the day, into classes for the study of various matters of useful knowledge, such as Geography, Arithmetic, Mensuration, &c., in their vernacular Hindí. The various Sanskrit works rendered into English for the other department are to be rendered also into Hindí for the use of these students; and some progress has been made in this direction.

Such,—as regards the enlisting of the Sanskrit on the side of progress,—is an outline of what has been attempted, during the last five years, in the Benares College. It was scarcely to be wondered at that a line of operations such as I have sketched,—designed to make the learning of the Hindús a strong ally instead of a stubborn opponent,—should occasionally be mistaken for a fond admiration of Sanskrit scholarship and an injudicious undervaluing of everything else in comparison. To show that this is quite a misconception, enough has been now said.

Benares College,
  31st July, 1851.   }   J. R. B.

THE FOLLOWING IS A TRANSLATION OF THE SANSKRIT PREFACE.

That intelligent persons should differ in opinion on any matter of importance is cause of regret. It is another cause of regret that intelligent persons should imagine that they differ in opinion where they really do not differ. How much of useless altercation or of unjust censure is likely to result from such mutual misapprehension, it would be superfluous to insist upon. It can never be an unworthy aim to endeavour to remove any mutual misapprehension which tends to make even two individuals think unworthily of one another : how much less so, then, where the mutual misapprehension exists between two great sections of the earth's inhabitants. The attempt is one in which even to have failed may be no disgrace.

The Hindús have an ancient literature which they cherish with affectionate pride. In this they do well. Europeans have a literature for which the praise is claimed that it is daily adding to the intellectual wealth of mankind. If Europeans regard their literature, in this aspect, with an affectionate pride, they also do well. But if the European, not rightly perceiving how the Hindú is asserting some unquestionable truth, in phraseology to which the European is unaccustomed, shall condemn the Hindú as asserting an error, he unknowingly commits injustice ; and the Hindú is liable to commit the like injustice in return. This liability to mutual misapprehension being very undesirable, it is desirable that a remedy for it should be devised.

As two opposite opinions cannot both be right, and as there can be no good, and there may be much harm, in holding a wrong opinion, it is reasonable that intelligent men, when they differ in opinion, should amicably discuss their difference. This, however, cannot be satisfactorily done unless they clearly apprehend what their difference amounts to ; and, as a difference commences where agreement ends, it is desirable in every case to determine as accurately as possible how far the agreement extends, in order that we may not be liable, in the heat of argument, to dispute about that which is not a matter of disagreement.

Having resolved, therefore, to attempt to determine the extent of agreement between European and Indian thinkers, it became a question what exposition should be assumed as most fully and fairly representing the sentiments of the Hindús. There was no difficulty in selecting the Nyáya in the first instance, because this system of philosophy, while it adverts, in some way or other, to all the topics considered in the other systems, takes cognizance also of various matters which the other systems disregard. Although other systems of philosophy, besides the Nyáya, are accepted in India, yet it suffices if we determine how far the sentiments of Europeans accord with those of the Nyaya, because, if a reader knows what relation the Nyáya bears to the Sánkhya or the Vedánta, with which he holds the Nyáya to be reconcileable, he must understand the relation of European philosophy to each of the other two if he understands its relation to this one. He who already knows the relative bearings of the three great places of pilgrimage, *Kásí*, *Gayá*, and *Prayága*, should be able to estimate the bearing of the second and third of these in respect of any fourth city, if he knows how that city lies in respect of the first.

In the first book of his Aphorisms, *GAUTAMA* lays down the plan of the whole Nyáya system. The reader to whom the present work is addressed is supposed to be conversant with the first book of the Nyáya Aphorisms. Such a reader will at once see how far we agree with GAUTAMA, and in regard to what points we differ from him. It has been thought advisable that, after determining thus much, we should then, instead of discussing the points of difference, rather go on to the development of the points of agreement, in the hope that, in the course of such an enquiry, the points of difference may either disappear or may have such light thrown upon them that we may be the better prepared to deal with them.

It has been the endeavour throughout to make the exposition as simple as possible, but the reader who is unacquainted with European literature must not be surprised if he meets here some things hard to understand. Almost every treatise on any novel topic requires to have light thrown upon it by an oral instructor. The reader, therefore, where he does not clearly see the sense, will do well to enquire of some one competent to explain and illustrate, rather than lay aside the book with the conviction that it is unmeaning because the meaning is not at once everywhere clear to him. It is in particular difficult to convey, by means of words alone, an adequate conception of such facts as can be best exhibited by experimental illustration.

The reader, if he be, as is not improbable, fond of abstract speculation, such as is found in the logical sections of the Nyáyä, ought not to regard with impatience the discussion of facts in regard to the properties of material things, to which we have seen fit to devote the 2nd Book, because, unless we have an ample body of recognised facts, it is impossible properly to exemplify and justify the logical processes of scientific enquiry which will furnish the subject of a future Book. If, on a hasty glance, it should seem to him that the discussion of wheels and hammers and ropes is beneath the dignity of philosophy, we would remind him that his own revered teachers of Astronomy give to the learner his first conceptions of the arrangement of the heavens by means of an armillary sphere formed of little hoops of bamboo, and that the Indian logician has not disdained to study the wheel of the potter, though he has not yet learned from it all that it is qualified to teach.

A

SYNOPSIS OF SCIENCE.

Glory be to God the Most High!

Having long pondered the import of the Aphorisms of Gautama, I compose a treatise, on a new plan, showing to what extent the sentiments of the English coincide with those of Gautama. Through the divine favour, may this work of mine, the result of not a little labour, treating of many matters in few words and yet plainly, prove acceptable to the intelligent.*

It is agreed alike by the learned of India and by the learned of other countries that the Chief End of Man† is not to be attained without a knowledge of the truth in regard to our souls and other things which it is desirable should be rightly known. Those, therefore, who aim at the Chief End of Man are bound to strive after the attainment of right knowledge.

The attainment of knowledge pre-supposes the possession of instruments adapted to its attainment; therefore, proceeding to

---

* As all Sanskrit books commence with an invocation or a benediction, suggesting in some measure the subject of the work, the example has been followed in the present instance.

† This expression of the Westminster catechism corresponds verbatim with the Sanskrit expression *parama-purush-ártha*.

consider these instruments, we give them the first place in the enumeration, here following, of the topics which we propose to consider in succession.

BOOK I.

SECTION I.

THE END OF THE PROPOSED ENQUIRY.

Aphorism I.

Since a knowledge of the truth is conducive to the attainment of the Chief End of Man, we propose to enquire into the truth respecting (1) our instruments for acquiring knowledge, (2) the objects to which these are applicable, (3) the state antecedent to knowledge, viz., doubt, (4) the motives for passing from the state of doubt to that of knowledge, (5) data to start with, whether popular or (6) scientific, (7) the steps in explicating from data what is implied in them, (8) the method of confuting error, so as to arrive at (9) certainty, (10) the nature of fair debate, (11) of wrangling, (12) of cavilling, (13) of fallacies, (14) of disingenuous artifices, (15) futile oppositions and (16) unfitness to be reasoned with.*

a. A Hindú reader will expect that we should here state formally the replies to four questions,—viz.—What is the objectmatter of the work?—What is the motive for reading it?—What

---

* In our first aphorism we depart from Gautama thus far, that whereas he declares the attainment of the Chief End of Man to result necessarily from a knowledge of the truth, we on the other hand regard knowledge as merely conducing to its attainment. What may be requisite in addition does not here fall to be considered.

is the relation between the object-matter and the work ?—and—
Who is the reader to whom the work is addressed ? To these
questions we thus reply :—The object-matter is whatever truth
we have got to tell in regard to the topics above enumerated ;—
the motive for reading it is supplied by the knowledge of truth
that may be found in it,—knowledge of truth being presumed
conducive to the attainment of the Chief End of Man ;—the relation between the object-matter and the work is that of a communication and its communicator ;—and the reader to whom the
work is addressed is he who, being devoid of malice, wishes to
know whatever of truth we have got to tell him.

b. As the knowledge of truth is not directly, but mediately,
conducive to the attainment of the Chief End of Man, the next
aphorism declares the order of the steps through which it conduces to that end.

Aphorism II.

The Chief End of Man is to be attained not without the removal of such False Notions as lead to those Emotions from which,
in our present Life, arise Actions such as lead to misery.*

a. What we mean to say is this, that there do exist erroneous
notions of various kinds ; and these give rise to improper feelings, such as covetousness or envy ; and these lead, in this life,
to wicked actions ; and these in their turn lead to misery. Now

---

* In this aphorism whilst retaining, for future reference, or substituting
equivalents for, the terms in Gautama's 2d aphorism, we dissent from him in
so far as we do not hold that all emotions deserve the name of faults (*dosha*),
nor that a man's being born into the world is the consequence of his actions
in any previous life, nor that the removal of false notions will necessarily
and of itself lead to beatitude. We agree with him only thus far that the
removal of false notions is conducive to the end desired, so far forth as their
existence places obstacles in the way of its attainment.

true knowledge puts an end to erroneous notions ; and, unless these depart, the improper feelings which they produce will not depart ;* and, unless these bad feelings be removed, the wicked actions which, in this life, arise from them, will not depart ; and unless wicked actions be done away with, the misery which these entail will not cease ; and the Chief End of Man is incompatible with the continuance of misery.

b. Now, since a definition will be looked for, of each of the things enunciated in the 1st aphorism, in the order of enunciation, we proceed to define and to divide the instruments in the acquisition of knowledge—these being what were first enounced.

<hr />

SECTION II.

THE INSTRUMENTS AVAILABLE IN PROSECUTING THE ENQUIRY.

Aphorism III.

Our instruments adapted to the attainment of right knowledge are (1) the deliverances of sense, and (2) the recognition of signs.

a. The name of a *proof* is also given to what is here termed an instrument adapted to the attainment of right knowledge or knowledge of the truth. We divide what is so termed into

<hr />

* It will be observed that whilst we hold that evil deeds tend, in the dispensations of Providence, to the evil of the doer, and evil deeds arise from evil inclinations, yet we do not say that all evil inclinations arise from erroneous notions. Whether that be the case or not, we are not now enquiring. We agree with Gautama only thus far—that whatever evils are due to a false notion may be expected to continue until the removal of that false notion in which they take their rise.

two—the same to which are usually given the names of Perception and Inference.

b. But then some one may ask,—Since Testimony and Comparison also are instruments by means of which we acquire right knowledge, how comes it that you say there are only two* such instruments ? To this we reply,—that Testimony is not a separately co-ordinate instrument of knowledge, because it is just the instrument of an inference (or, in other words, Testimony, when received in evidence, is " the recognition of a sign") where the sign is an assertion. So again, Comparison (or the ascertaining that an object never previously beheld by us is so and so, because it resembles some well-known object which we had been told that so and so resembled,) is nothing else than inferring—in reliance on testimony. Hence we hold that there are only two co-ordinate divisions of proof.

c. We proceed to define the instruments of knowledge, as thus bi-partitely divided, in their order.

Aphorism IV.

By a deliverance of sense is meant the knowledge which has arisen from the contact of an organ of sense with its object.

a. That is to say,—that instrument of correct knowledge which arises from the contact of an organ of sense with its object is a deliverance of sense, or a sensation—such as that of Sight for example.

---

* Gautama holds that we have four instruments for acquiring knowledge. Kapila, the founder of the *Sánkhya* school does not recognise Comparison as co-ordinate with the others ; and *Kanáda*, the founder of the *Vais eshika* school, recognises only two members in the first division, giving the reason which we have adopted above.

b. Next we have to define and divide the ' recognition of a sign.'

Aphorism V.

Now the recognition of a sign, which is preceded thereby, is of three kinds, (1) having (as the sign) the *prior,* or (2) having (as the sign) the *posterior,* or else (3) consisting in the *perception of homogeneousness.*

a. In saying ' preceded thereby' we mean ' preceded by a deliverance of sense' (—for else the recognition of the sign were impossible).

b. The inference is said to be *à priori* (or, in other words, the instrument of our knowledge involves something *prior* to what is thereby known,) when an effect is inferred from its cause ; as when, from the rising of clouds, it is inferred that there will be rain.

c. The inference is said to be *à posteriori* (or, in other words, the instrument of our knowledge involves something *posterior* to what is thereby known,) when a cause is inferred from its effect ; as when, on seeing that the water of a river is changed in colour from what it was before, or that the channel has become full, or that the stream runs more rapidly, we infer that there has been rain.

d. The inference is said to be *from analogy* (or, in other words, ' the perception of homogeneousness' is the instrument of our knowledge,) when, for example, on seeing somewhere a mango-tree in blossom, we infer that the mango-trees in other places also are blossoming.

ε. Although we do not hold Comparison and Testimony to be, as proofs, co-ordinate with the foregoing, we shall follow Gautama's order, in now describing them, so that the reader may know what cases of inference these two are. First, then, Comparison, as defined by Gautama, is as follows.

Aphorism VI.

The 'recognition of likeness' is the instrument in the ascertaining, of that which was to be ascertained, through similarity to something previously well-known.

a. That is to say,—as when a man, having been told that the animal called the Bos Gavaeus is like a cow, recognises the Bos Gavaeus, on first seeing it, from its likeness to a cow, so in other cases where certainty is arrived at through the recognition of the resemblance of anything to something previously well-known, the instrument of knowledge, according to Gautama, is Comparison (*upamána*).

b. Testimony is defined by Gautama as follows.

Aphorism VII.

Testimony (as an instrument of *right* knowledge) is the word of one worthy (to have his words implicitly accepted as an authority).

a. Testimony is divisible as follows.

Aphorism VIII.

It is of two kinds, in respect that it may be ' that whereof the matter is seen', or ' that whereof the matter is unseen.'

a. By the ' it' is meant testimony. By the testimony, or verbal expression, ' whereof the matter is seen' is meant that speech the thing referred to in which is here patent to the senses,—as when the river Ganges is what is spoken of. That speech, the things referred to in which are not here patent to the senses, is what we mean by ' that whereof the matter is unseen,'—as is the case when Paradise or the like is spoken of.*

b. Here, for the present, we quit the topic of the instruments available for the attainment of correct knowledge.

c. As it will be expected that we should next consider what it is to which these instruments are applicable, we now turn to the topic of the *objects* of cognition.

SECTION III.

THE OBJECTS ABOUT WHICH THE ENQUIRY IS CONCERNED.

d. We divide and define as follows the objects concerning which it is desirable that we should have right notions.

Aphorism IX.

Soul, body, sense, sense-object, understanding, the mind, evil deeds, evil passions, mundane life, retribution, pain, and the Chief End of Man,—such are the objects concerning which it is desirable that we should have right notions.†

a. Among these objects, the one first enounced, viz. Soul, may be thus defined.

---

* In other words, assertions are of two kinds—where the thing asserted is open to verification by the senses, and where it is not.

† Here we depart from Gautama in substituting ' evil deeds' for ' activity' (which, as before remarked, we do not allow to be essentially faulty),—' evil passions' for ' faults,' under which head Gautama lumps all the emotions, good or bad,—and ' mundane life' for ' transmigration,' of which more anon.

Aphorism X.

Desire, aversion, volition, pleasure, pain, and knowledge, are the sign of the Soul.

a. The word ' sign' here means a characteristic mark—or that whereby such and such an object is recognised.

b. Next in order we have to define the sense of the term Body.

Aphorism XI.

The body is the site of muscular action, of the organs [of sense], and of the sentiments [of pleasure or pain experienced by the soul].

a. ' Muscular action' (*cheshṭá*) is that species of action which originates in voluntary effort (*prayatna*).

b. The ' organs [of sense]' are the Sight &c.

c. The ' sentiments [here spoken of]' are pleasure and pain.

d. The meaning is this, that the body is the site of all these ; and that each of these severally, viz., the being the site of muscular action, &c., is a characteristic of the body.

e. Now we have to divide and define the ' organs,' which next present themselves.

Aphorism XII.

The organs [of sense], viz., the Smell, the Taste, the Sight, the Touch, and the Hearing, are what apprehend the qualities of the Elements and of things formed of these.

a. Which are those ' Elements' shall be considered presently. We mean here, when we say that the organs of sense are what apprehend the qualities of these external things, that

B

each organ is distinguished by its apprehension of its own particular object [—the Sight apprehending colour, the Hearing having sound for its object, &c.].

b. As we shall be asked what we mean by the 'Elements'
we reply :—

Aphorism XIII.

The Elements are Gold &c.*

a. Whether 'earth &c.,' are entitled to the name of 'Elements
(*bhúta*) we shall consider in the sequel.

b. We have now to divide and define ' Sense-object,' which
presents itself next in order.

Aphorism XIV.

Their objects are the qualities of the Elements and of things
formed of these,—meaning the qualities odour, savour, colour,
tangibility, and sound.

a. By the 'Elements' we mean Gold &c., meaning by the
' &c.,' various things that will be spoken of in the sequel. The
qualities, here referred to, of these various objects, are odours
&c.,—these being the objects of the senses.

b. We proceed to define ' understanding.'

---

* Here we are constrained to forego the employment of the language of
Gautama, who says that " Earth, Water, Light, Air, Ether,—these are the
' Elements' *(bhúta)*." What may be said for and against the theory that there
are five special substrata for the qualities which affect the five senses, we may
have occasion to consider further on. In the mean time, to denote 'external
things we may employ the term *bhútádi*, i. e., ' the elements &c.,' meaning
by the   &c.' whatever is compounded of the elements.

Aphorism XV.

Understanding (*buddhi*), apprehension (*upalabdhi*), knowledge (*jnána*),—these terms are not different in meaning.

a. ' Not different in meaning'—i. e. synonymous.

b. ' Next we have to characterise the Mind.

Aphorism XVI.

The characteristic of the Mind is this, that there does not arise [in a single Soul] more than one cognition at once.

a. ' At once'—i. e., simultaneously. Of course you must supply ' in a single soul.'

b. The sense of the Aphorism is this, that the sign, or characteristic, whereby we recognise the mind (or the internal organ of the Soul), is the habit in virtue of which more thoughts than one do not occur at once.*

c. We have next to define Activity.

Aphorism XVII.

Activity is that which originates the [cognitions of the] understanding and the [gestures of the] body.

---

* The English reader who is accustomed to hear the words Soul and Mind employed interchangeably must avoid all such laxness of phraseology when speaking with a Hindú metaphysician. We shall have occasion in the sequel to compare the opinion of Gautama with that of Dugald Stewart respecting the non-simultaneousness of cognitions. Meantime let it be carefully remembered that the cognizant principle is here spoken of as the Soul, and that the term Mind (*manas*) stands for the ' internal organ,' or faculty, which ministers to the Soul—or Self—as the ' external organs' also minister to it.

a. Thus activity (*pravritti*) is of two* sorts, through its setting the Understanding to work or the Body.

b. It is also divisible into two kinds through the distinction of proper and improper.†

c. We have now to define the Passions.

Aphorism XVIII.

[The Passions] Desire &c. have this characteristic, that they actuate.‡

a. ' That they actuate'—i. e., that they are causes of Activity. Desire &c., are what are distinguished by being such. The meaning of the ' &c.' will be told in the sequel.

b. We have now to define mundane life.

Aphorism XIX.

[Mundane] life is the union of the soul with such bodies as we have here.§

---

* Gautama makes it of three sorts, holding that to be a separate kind which originates the utterances of the voice. This we look upon as being included under the second head.

† With this distinction Gautama, for reasons given in our version of the first book of his Aphorisms, is not concerned. Not agreeing with him that *all* Activity is to be eschewed, we are called upon to mark the distinction. Of this more anon.

‡ Gautama, holding that all Activity ought to be eschewed, gives to all the passions, good and bad alike, the common name of ' faults' or ' failings' (*dosha*). Not being Quietists, we cannot follow him in this. We have modified the Aphorism accordingly.

§ Gautama defines life as a state of transmigration. This view we do not adopt. Our own definition is not at variance with Gautama's view, although we demur to the proposition that every one who is now born has been born again and again.

a. We have next to define the fruit of actions, or Retribution.

Aphorism XX.

The fruits of our actions are those things [whether pleasurable or painful] which are the consequences of our Activity.

a. That is to say, retribution is the experiencing of the pleasures or the pains which result from Activity properly or improperly directed.

b. Pain is next to be defined.

Aphorism XXI.

Pain is that which is in the shape of vexation.

a. That is to say, pain is that which is recognised by distress or annoyance.

b. We proceed next [not to define but] to show the Chief End of Man in conjunction with its cause.

Aphorism XXII.

The Chief End of Man is to be attained through the grace of God.

a Here we quit the topic of the definition of the things concerning which it is desirable that we should entertain right notions.

b. Now the bare enunciation and definition of the Objects,

---

* Gautama defines the Chief End of Man to consist in absolute emancipation from pain and joy alike. For the discussion which that definition is calculated to provoke, this is not the place.

concerning which it is desirable that we should entertain right notions, does not ensure a correct knowledge and belief of them. Demonstration may be needed; and this is preceded by Doubt. We have now therefore to define the state intermediate between hearing and believing—or Doubt.

SECTION IV.

COMPLETING THE TOPIC OF THE PRE-REQUISITES OF REASONING.

Aphorism XXIII.

Doubt is [what results] from the perception of a sameness, [conjointly with] the non-perception of a difference, and the remembrance of a difference.*

a. That is to say,—Doubt arises 'from the perception of a sameness,'—i. e., from the apprehending of some concrete object which possesses characters common to it with other objects ; and ' from the non-perception of a difference'—i. e. from the not discerning any character such as distinguishes the object from all other objects ;—and 'from the remembrance of a difference,'—i. e., from the taking into consideration some special character—such as the fact of being a post or the fact of being a man—[the remembrance of which alternatives keeps up the doubt so long as there is nothing to enable us to decide, in the dusk for instance, whether the object which appears to be of the height of a man is a man or a post].

---

* We have taken this definition from *Kanáda* (2d Lecture), because it is simpler than that given by Gautama, in regard to the interpretation of whose 23d Aphorism (—See the English version—) the commentators are not agreed.

b. Now, since one makes no effort for the removal of doubt in the absence of a motive, we have next to define a Motive.

Aphorism XXIV.

That thing which, when placed before us, causes us to act, is a Motive.

a. 'Placed before us'—i. e. proposed to us. So the meaning is this, that a motive is an object of desire, which [— whether the desire be to obtain it or to escape it—] is the cause of our acting.

b. But, in every enquiry, to reach the unknown we must start from the *known ;*—there must be *data.* The knowledge which, in any enquiry, we may treat as requiring no demonstration, is either popular—being that on which the unlearned and the learned are at one ;—or it is scientific—belonging to the schools. First then we have to define a popular datum or ' familiar case of a fact.'

Aphorism XXV.

In regard to [some fact respecting] what thing both the ordinary man and the acute investigator entertain a sameness of opinion, that [thing] is called a ' familiar case' [of the fact in question].

a. Here, by ' the ordinary man' we mean the one who stands in need of instruction, and by the ' acute investigator' one competent to instruct him. Whatever thing they entertain a ' sameness of opinion'—i. e., no difference of opinion—regarding, that thing may serve as a ' familiar case' (*drishtánta*) in regard to any fact that may be under discussion. So, what we mean to be understood by a ' familiar case of a fact' is anything in regard to which both the parties in a dispute are agreed that there is no doubt,—

as, for example, in proving that there is fire in such and such a place, the *culinary hearth* will be accepted by every one [in India] as a familiar case of a locality where there is always fire ; and in proving that there is *not* fire in such and such a place, a *deep lake* will be admitted to be a familiar case of a locality where fire is *not* to be met with.

b. Here closes the topic of the pre-requisites of reasoning [—for, although scientific data—see Aph. 24. *b.*—are available, in the schools, as premises, yet they were first established by a process of argumentation].

c. We have next to define a scientific ' tenet.'

SECTION V.

OF PROPOSITIONS, NOT FAMILIAR, THAT MAY BE EMPLOYED IN REASONING WITHOUT REQUIRING TO BE EACH TIME DEMONSTRATED.

Aphorism XXVI.

A ' tenet' is that, the steadfastness of the acceptance of which rests on a treatise [of weight and authority].

a. That is to say,—a tenet is an assured conviction in respect of some matter which has been determined in some system.

b. We have next to divide [the ' tenets' thus characterised generally].

Aphorism XXVII.

[Tenets are divided into the species that are described in the succeeding aphorisms] through the difference between a ' Dogma of all the schools,' a ' Dogma peculiar to some one or more schools,' a ' Hypothetical Dogma,' and an 'Implied Dogma.'

a. Thus scientific tenets are of four kinds. First then we have to define a ' Dogma of all the schools' (*sarva-tantra-siddhánta*).

Aphorism XXVIII.

That [position or tenet] which is not in opposition to any of the schools, and which is advanced [as a tenet] by [at least] some one school, is [what we mean by] a ' Dogma of all the schools.'

a. Such a tenet of all the schools, or of all systems of doctrine, is the fact that *odour* is to be apprehended by the sense of *smell*, and so on.

b. We have next to define a ' Dogma peculiar to some school' (*prati-tantra-siddhánta*).

Aphorism XXIX.

That [position] which is [held] established in the same school, and which in another school is [regarded as] not established, is [what we mean by] a ' Dogma peculiar to some school.'

a. By ' established in the *same* school' we mean established in *one* school and not in another. A dogma is said to be of this kind—or sectarial, not catholic—when it is accepted by only one of the parties in a dispute. Such, for example, is [at present] the European tenet of the earth's motion.

b. The next to be defined is the ' Hypothetical Dogma' (*adhikarana-siddhánta*).

Aphorism XXX.

That, if which be [held] established, [—and not otherwise—]

C

there is the establishing of another point, is [what we mean by] a ' Hypothetical Dogma.'

a. The meaning is this—that, that position [—for which no evidence is offered in the first instance—] is a hypothetical dogma [or hypothesis] only on the establishment of which taking place [—by its being conceded—] does the establishment take place of another proposition under consideration. For example,—you can establish that God is *omniscient* only when you have established that the world was *produced* [—for unless this be conceded, it is in vain to argue that the world furnishes any evidence of the omniscience of its *producer*].

b. Lastly we have to define an 'Implied Dogma' (*abhyupaga-ma-siddhánta*).

Aphorism XXXI.

The mention of some particular character [which mention, by the leader of a school, could have arisen only] from his having held some opinion which he has not anywhere explicitly declared, determines an ' Implied Dogma.'

a. That is to say,—an 'Implied Dogma' is determined by the mention of some particular fact, by which we are made aware that the teacher held some opinion which he did not explicitly declare. Thus, for example, Gautama [by speaking of the Mind as one of the *instruments* of knowledge] implies that he reckons the Mind among the *organs* [of the Soul, although he nowhere explicitly lays down this tenet, which is a tenet not held, for instance, by the followers of the *Sánkhya* school].

Here ends the topic of the definition of ' tenets' that take their place in argumentation.

We have next to divide—previously to defining—the Members [of a demonstration] which present themselves next in order [of the topics enumerated in the first aphorism].

SECTION VI.

THE METHOD OF ARGUMENTATIVE EXPOSITION.

Aphorism XXXII.

The members [of a demonstration] are (1) the Proposition, (2) the Reason, (3) the Example, (4) the Application, and (5) the Conclusion.

a. Now we have to define the Proposition (*pratijña*).

Aphorism XXXIII.

The Proposition is the declaration of what is to be established.

a. That is to say,—the member [of an argumentative exposition] in which we state what we intend to prove, is the Proposition. For example—" The Earth is in the form of a globe," is a Proposition [—what we here undertake to prove of the Earth being its *globularity*].

b. We shall next define, and, by [reference to] the two subsequent aphorisms, divide, the Reason (*hetu*), which presents itself next in order [of those enumerated in Aphorism 32].

Aphorism XXXIV.

The Reason is the means for the establishing of what is to be established ; [and this force it may derive either] from the Example's having a character which involves another, or [conversely] through the Example's wanting a character the want of which involves the absence of another.

a. Here the generic definition is this—that ' the Reason is the means for the establishing of what is to be established.' That a Reason may be of *two* kinds, is declared in the assertion that it derives its force as an argument, (1) from the Example's having a character which involves another, or (2) from the Example's wanting a character the want of which involves the absence of another.

b. The ' having a character which involves another' (*sádharmya*) is what is also spoken of as ' Agreement' (*anwaya*,)—meaning thereby ' the fact of some positive character's being invariably attended by some other.' Thus then the Member [of an argumentative exposition] which [Member, or Minor Premiss, as well as the Reason which it sets forth] is also technically spoken of as ' the Reason', may be that which propounds a Reason which stands in relation to a Universal Affirmative. For example, [we may assign as a Reason for the Proposition that ' The Earth is a globe'] the assertion—' Because its shadow [on the Moon] is [invariably] circular'—and [this we may allege as standing in relation to the Universal Affirmative that] ' whatever has an invariably circular shadow is seen to be a globe,—as is the case with a billiard-ball, for example.'

c. The ' wanting a character the want of which involves the absence of another' (*vaidharmya*) is what is also spoken of as ' Disagreement' (*vyatireka*),—meaning thereby ' the fact of some given negative character's being invariably attended by the negation of some other character.' Thus then the Member termed ' the Reason' [or Minor Premiss] may be that which propounds a Reason which stands in relation to a Universal Negative. For example, we may assign as a Reason, in support of the foregoing Proposition, the same assertion as before—viz., ' Because its shadow is invariably circular,' and [this we may allege as

standing in relation not as before to a Universal Affirmative but
to the following Universal Negative, viz.,] "Nothing which is
not a globe casts invariably a circular shadow,"—as a post, for
instance.

d. We have now to define the ' Example' which presents itself next in order [of those enumerated under Aph. 32].

Aphorism XXXV.

The Example is some [undisputed] 'familiar case of a fact'
[see Aph. 25], which, through its having a character which is
invariably attended by what is to be established, establishes, [in
conjunction with the Reason] the existence of that character
which is to be established.

a. When we employ the term 'Example' to denote that third
Member of a demonstrative exposition in which the actual Example is propounded, we mean by the term ' Example' that [Ma-
jor] Premiss which causes us to admit the Subject's possession
of what is to be proved, this being seen to follow from the possession of the instrumental mean [which we have already spoken
of as the 'Reason']. For example—when it is alleged that every
thing which invariably casts a circular shadow is globular—' as
is a billiard ball,' [which invariably casts a circular shadow,—
the recognition of this fact causes us to recognise the globularity
of the Earth—the Earth having been already allowed to possess
the character of invariably casting a circular shadow].

b. [But the 'Example,' as already remarked, may be of two
kinds; so that] we have now to define the Example in which
some two given characters [instead of being both present] are
both absent.

Aphorism XXXVI.

Or the Example, on the other hand, wanting some character
the want of which involves the absence of some given character,

[may co-operate in establishing what is to be established] by a process the converse* of that [declared in the preceding aphorism].

a. 'By a process the converse of that'—i. e. by exhibiting the invariable absence of the alleged Reason where there is the absence of what is to be established. And thus an 'Example of concomitant negatives' (*vyatirekyudáharana*) means one which [—conveyed in a Major Premiss consisting of a Universal Negative—] declares the invariable absence of the alleged Reason where there is the absence of the character the presence of which we wish to establish. For instance—'Nothing which is not a globe casts invariably a circular shadow,—as a post, for example, [—or as a circular disc, which, not being a globe, does not invariably cast a circular shadow, though it may do so in certain positions].

b. We have now to define the 'Application' (*upanaya*) which presents itself next in order [of those enumerated in Aphorism 32].

Aphorism XXXVII.

The 'Application' is the re-statement of that in respect of which something is to be established,—this re-statement declaring it to be *so* or *not so,* as regards the 'Example.'

a. By 'that in respect of which something is to be established,' we here mean the Subject of the Proposition. By the 're-statement as regards the Example,' we mean that Member [of an

---

* This refers to the " conversion by *negation ;* or, as it is commonly called, by *contra-position*," of which Whately speaks in his Logic, B. II. ch. 2d. § 4., remarking that every Universal Affirmative may " be fairly converted in this way" *e. g.,* Every poet is a man of genius; *therefore* He who is not a man of genius is not a poet :" (or, " None but a man of genius can be a poet").

argumentative exposition] in which the Subject is declared to agree with the Example, or to differ from it, as far as regards the question in hand. This re-statement may assume two forms according as the Example is direct, or negatively converse. The re-statement of the Subject takes the form of—' And this is so' —when the re-statement has reference to a direct Example. Thus the re-statement may be 'And this Earth is so, [—when the Major Premiss involves a direct Example,—i. e., when it runs thus—' Whatever is possessed of a shadow invariably circular is a globe—as a billiard-ball is']. The re-statement in regard to the Subject takes the form of ' And this is *not* so,'—when the re-statement has reference to a negatively converse Example. [Thus the re-statement may be ' And this Earth is *not* so'—when the Major Premiss involves a negatively converse Example ;—i. e., when it runs thus—' whatever is not globular is not possessed of a shadow invariably circular,—as a post, for example ;'—but 'this Earth is *not* so,' for the Earth *has* a shadow invariably circular].

b. We have now to define the ' Conclusion' (*nigamana*).

Aphorism XXXVIII.

The ' Conclusion' is the re-statement of the ' Proposition' [as being now authorized] by the mention of the ' Reason.'

a. The ' mention of the Reason'—i. e. the mention [—involved in the steps between the Proposition and the Conclusion—] of a character of the Subject which is distinguished by being invariably attended [by that which we wish to establish]. This re-statement of the Proposition—authorized thereby—constitutes the Member [of an argumentative exposition] which we mean by the ' Conclusion.' For example, [our conclusion may be] ' Therefore [—i. e. for that Reason—] the Earth is a globe.'

b. Here concludes the topic of the method of argumentative exposition.

c. But, thus far, we have been shown an arrangement for hearing only one side of the question, and how can we be sure that the opposite is not the right one ? Before making up our minds we must hear both sides. Next, therefore, before defining ' Ascertainment' we have to define the ' Confutation' (*tarka*) of objections.

SECTION VII.

CONCLUDING THE TOPIC OF DEMONSTRATION.

Aphorism XXXIX.

' Confutation'— [which is intended] for the ascertaining of the truth in regard to any thing the truth in regard to which is not thoroughly discerned—is reasoning from the presence of the reason [which would not be present if that which is to be established were not present].

a. [In other words, the confutation of him who denies the conclusion of a sound argument while he admits the premises, consists in our directing him to look at it from an opposite point of view. To one who admits that the Earth's shadow on the Moon is invariably circular, and that what casts a shadow invariably circular must be a globe, but who still hesitates to admit that the Earth is a globe, we remark]—for example—' If the Earth were *not* a globe, it would not have a shadow invariably circular.'

We have now to define ' Ascertainment' (*nirṇaya*) which presents itself next in order.

Aphorism XL.

'Ascertainment' is the determination of a matter by dealing with both sides of the question, after having been in doubt.

a. That is to say—'Ascertainment' is the settling of the question by the establishing of the one view of it and the confuting of the other view.

b. Here ends the first diurnal portion of the first Book.

c But an honest enquirer after truth may have heard both sides and still be in perplexity. Amicable ' Discussion' (*váda)* is then open to him ; and this we proceed to define.

THE SECOND DIURNAL PORTION.

SECTION VIII.

THE TOPIC OF CONTROVERSY.

We have now to define amicable ' Discussion' (*váda*).

Aphorism XLI.

' Discussion' is the undertaking [—by two parties respectively—] of the one side and of the other in regard to what [conclusion] has been arrived at by means of the five-membered [process of demonstration already explained;—this discussion] consisting in the defending [of the proposition] by proofs [on the part of the one disputant] and the assailing it by objections [on the part of the other,—the discussion being conducted on both sides] without discordance in respect of the tenets [or principles on which the conclusion is to depend].*

a.   Since those who, (under the pretence of seeking the truth,) are dishonestly eager for victory, will *wrangle*, we have next to define ' Wrangling' (*jalpa*).

Aphorism XLII.

' Wrangling,' consisting in the defence or attack [of a propo· sition] by means of ' Frauds' [—see *Aph.* 50—], ' Futilities' [—see *Aph.* 58—], and ' What calls for nothing save an indignant rebuke' [—see *Aph.* 59.—], is what takes place after the procedure aforesaid [—that is to say after a fair course of argu· mentation,—supposing this to have failed to bring the disputants to an agreement].

a.   So—what we mean by ' Wrangling' is the discourse of one bent only on victory, and indifferent which side of the question is established, provided he secure the credit of a victory.

b.   Disingenuous persons who have not sufficient skill for ' Wrangling' [—which implies the possession of sufficient skill to take up a position and maintain it—], have recourse to *ca-villing.* We have next, therefore, to define ' Cavilling' (*vitandá.*)

Aphorism XLIII.

This [—viz., ' wrangling'—], when devoid of [any attempt made for] the establishing of the opposite side of the question, is ' Cavilling.'

---

* The reader who collates the English version with the Sanskrit will observe that here, as in various other places, explanatory clauses are inserted in the English version of the Aphorism instead of being subjoined to the Aphorism, as is customary in Sanskrit works.

a. That is to say,—when a disingenuous disputant, eager for victory, attempts to establish nothing, but confines himself to carping at the arguments of the other party, he is said to cavil.

b. Here the topic of ' Controversy' is concluded.

c. Now, wranglers and cavillers, when they do not find good reasons to support their positions, make use of what merely *look* like reasons;—and even the honest enquirer, by mistake, may make use of such;—therefore we now procede to define and divide the ' Semblance of a reason' (*hetwábhásá*) which presents itself next in order [of the topics enumerated in the first Aphorism].

SECTION IX.

OF FALLACIES, OR WHAT ONLY LOOK LIKE REASONS, BY MEANS OF WHICH A MAN MAY DECEIVE HIMSELF OR ANOTHER.

Aphorism XLIV.

The ' Semblances of a reason' are (1) the 'Erratic,' (2) the ' Contradictory,' (3) the 'Equally available on both sides,' (4) that which is ' In the same case with what is to be proved, and (5) the ' Mistimed.'

a. By the ' Semblance of a reason' we mean a mere appearance which is *like* a reason, or, in short, a *bad* reason.

b. Of this class we proceed to define the first, viz., the ' Erratic' (*savyabhichára*).

Aphorism XLV.

That [semblance of a reason] is ' Erratic' which attends not merely the one [character, which it is employed to prove, but attends also the absence of that character].

a. This ' Erratic' semblance of a reason may be subdivided, as we shall have occasion to see in the sequel.

b. We have now to define the ' Contradictory' (*viruddha*) semblance of a reason, which comes next in order.

Aphorism XLVI.

That [semblance of a reason] is called ' Contradictory' which is repugnant to what is proposed to be established.

a. In other words,—after having proposed, or stated, that which is to be proved, a ' Contradictory' semblance of a reason is one employed which is opposed thereto, or invariably attended by the *negation* of what is to be established :—as if, for example, one were to argue " This is fiery,—because it is a body of water."

b. We have now to define that semblance of a reason which is ' Equally available on both sides' (*prakaranasama*),—this presenting itself next in order.

Aphorism XLVII.

That [pretended reason] from which a question may arise as to whether the case stands this way or the other way, if employed with the view of determining the state of the case, is ' equally available on both sides' [of the dispute].

a. That is to say ;—that reason employed, or adduced, for the ascertainment of one's own proposition, or the negation of the other's proposition, is called ' the same for both sides' :—but *which* reason ?—that from which two opposite views may arise [—so that nothing is settled by it at all].

b. For example,—suppose a man argues that Sound is eternal, because it has the nature of Sound, his opponent will be provoked to retort that Sound is *not* eternal, just because it has the nature of Sound,—the argument having just as much force on the one side as on the other.

c. We have now to define that semblance of a reason which is 'In the same case with what is to be proved' (*sádhyasama*),—this presenting itself next in order.

Aphorism XLVIII.

And it [—the alleged reason—] is ' In the same case with what is to be proved,' if, in standing itself in need of proof, it does not differ from that which is to be proved.

a. If the reason stands in need of being proved too, just as the proposition stands in need of being proved, then it is said to be in the same case with what is to be proved, and therefore the expression ' unestablished' (*asiddha*) is employed in speaking of such a reason. And this ' unestablishedness' or ' unreality' is of several descriptions, as will be seen in the sequel.

b. We have now to define the ' Mistimed' (*kálátíta*) semblance of a reason,—which comes next in order.

Aphorism XLIX.

That [semblance of a reason] is said to be ' Mistimed' which is adduced when the time is not [that when it might have availed].

a. For example,—suppose one argues that fire does not contain heat ' because it is factitious,'—the argument is ' Mistimed' if we have already ascertained, by the superior evidence of the senses, that fire [—granting it to be factitious—] *does* contain heat.

b. Here concludes the topic of the Semblance of a reason.

c. We remarked that an honest enquirer may, inadvertently, take the semblance of a reason for a real one ; but what we have next to define are the tricks which are employed only by the dishonest disputant.

SECTION X.

OF THE TRICKS EMPLOYED BY THE DISHONEST DISPUTANT TO THWART THE OTHER PARTY.

Aphorism L.

' Unfairness' (*chhala*) is the opposing of what is propounded by means of assuming a different sense [from that which the objector well knows the propounder intended his terms to convey].

a. We proceed to enumerate the divisions of ' Unfairness.'

Aphorism LI.

It [—viz., ' Unfairness'—] is of three kinds, (1) Fraud in respect of a term, (2) Fraud in respect of a genus, and (3) Fraud in respect of a trope.

a. Of these we shall now define ' Fraud in respect of a term' (*vák-chhala*).

Aphorism LII.

' Fraud in respect of a term' is the assigning a meaning other than [the objector well knows] was intended by the speaker when he named the thing by a term that happened to be ambiguous.

a. For example,—if some one says " A cow (*gau*) has horns,' —a caviller, recollecting that the word *gau* is explained in the dictionary to mean an elephant as well as a cow, may exclaim—

" What !—do you say that an *elephant* has horns ?"

b. We have next to define ' Fraud in respect of a genus'
(*sámányachchhala*).

Aphorism LIII.

' Fraud in respect of a genus' is the assuming that something
is spoken of, in respect whereof the fact asserted is impossible,
because [forsooth] this thing happens to be the same in *kind* with
that of which the fact asserted *is* possible.

a. For example, on some one's saying, ' This is a ' Bráhman,'
—he must be possessed of learning and conduct' ;—the other,
assuming that he here deduces the possession of learning and
conduct from the fact of being a Bráhman, says—' How can that
' be ?—for, the possession of learning and conduct, if deducible
' from the fact of being a Bráhman, would be found, where it
' cannot, in his *childhood.*' [The other, of course, meant, as the
objector very well knows, to speak of a Bráhman who has lived
long enough in the world to render it possible for him to study,
in which case the probability is that he will have studied].

b. We have next to define ' Fraud in respect of a trope'
(*upachárachchhala*).

Aphorism LIV.

' Fraud in respect of a trope' is the denial of the truth of the
matter, when the assertion was made in one or other of the
modes, [viz., literal or metaphorical,—which it suits the purpose
of the objector to invert].

a. For example, in the case of such an as sertion as ' The scaf-
folds cry out' [—somewhat analogous to the English phraseology
' The pit and gallery applauded'—] a dishonest opponent will say
' It is only *those standing* on the scaffolds that cry out' [—as if the

other had meant to make the assertion literally of the scaffolds.

b. Here it may be proper to anticipate an objection.

Aphorism LV.

' Fraud in respect of a trope' [some one may fancy at first sight—] is just ' Fraud in respect of a term,' for it does not differ therefrom.

a. The meaning of this objection is, that Fraud is of only two kinds, not of three kinds; for Fraud in respect of a trope is just Fraud in respect of a word, seeing that these agree in the assumption that some word was employed in another sense than that in which it was well enough known that the speaker did employ it.

b. This objection we proceed to overrule.

Aphorism LVI.

It is not so [as the objector in Aphorism 55 supposes,] because they *do* differ.

a. To show that the opposite view involves an absurdity we remark :—

Aphorism LVII.

Or if there were no distinction where there is *any* similarity of character [—as there no doubt is between the two species of Fraud under consideration—] then we should have only *one* kind of Fraud.

a. That is to say,—if no distinction is to result from any property whatever provided there be *some* similarity of nature, then ' Fraud', in as much as each variety has a common nature so far forth as each *is* a ' Fraud,' would not be even of *two* kinds, as the objector is understood to hold that it is.

b. Here concludes the topic of Fraud in disputation.

c. We have now to define 'Futility,' which presents itself next in order.

SECTION XI.

OF FUTILE OBJECTIONS AND HOPELESS STUPIDITY.

Aphorism LVIII.

Futility consists in the offering of objections founded on [some mere] similarity or difference of character [—without regard to the question whether the fact asserted bears any invariable relation to that character].

a. The expression 'founded on similarity or difference of character' is a definite one [—intended to convey just so much, and to exclude everything beyond—]; therefore the meaning is this, that Futility consists in objecting, or taking exception, on the ground of similarity or difference of character *without respect to invariableness of association* [between the character and that whereof it is taken as a sign of the presence or the absence. For example, if it were propounded that 'The man is unfit to travel, because he has a fever,' it would be futile to object that 'The man *is* fit to travel, because he is a soldier'—there being no invariableness of connection between the being a soldier and the being fit to travel].

b. We have now to define 'Unfitness to be argued with' (*nigrahasthána*),—the topic which presents itself next in order.

E

Aphorism LIX.

Unfitness to be argued with consists in one's [stupidly] mis-understanding, or *not* understanding at all.

a. The term here rendered ' Unfitness to be argued with' sig-nifies literally the place, i. e. the suggester, of censure or re-buke ; [—for if a man stupidly misunderstands you or does not understand you at all, and yet still persists in trying to make a show of opposition, then the matter has come to that point where there is nothing left for it but to rebuke him and drop the discus-sion].

b. In order to prevent the mistake, [into which some might fall, of supposing] that there is no subdivision of Futility and Unfitness to be reasoned with, [—the subdivisions of which will be stated in their proper place—] we remark as follows.

Aphorism LX.

Since each of them is of different kinds, ' Futility' and ' Unfitness to be reasoned with' are of various descriptions [which will be considered in the sequel].

a. But, as other questions are more pressing, their subdivision is not made at present ;—such is the import of the Aphorism.

END OF THE FIRST BOOK.

BOOK II.

SECTION I.

OF THE FIVE SENSES.

In the 1st Book* the Senses and their Objects were enumerated. Now each of these is to be described more fully ; and
first of the Sense of Smell† (*ghrána*).

Aphorism I.

The organ of Smell resides in the nose.

a. Particles issuing from odorous bodies and drawn into the

---

* In the second and subsequent Books we do not follow the order of
Gautama's second and subsequent Books, His first Book contains the germs
of his whole philosophy. Having re-written the first Book, we proceed to deve-
lope its germs in the order that seems most likely to lead on the understanding from the point reached to the point which we desire to reach.

† Reid (—see Sir Wm. Hamilton's edition, p. 104) says that the principle
of proceeding, in an exposition, from the more simple to the more complex,
" ought to determine us to make choice even among the senses, and to give
the precedence, not to the noblest or most useful, but to the simplest."
Gautama follows, to some extent, the same order as Reid, but for a different
reason, his arrangement of the five senses being determined by the order in
which he discusses his five Elements—his five substrata of the qualities apprehended by the senses respectively. Among his Elements, he gives the precedence to that which he holds to possess the greater number of sensible qualities.

nose in breathing, come into contact with the organ, and then, in the absence of preventing causes, the sensation of smell takes place.

b. Hence that body is called odorous in which there resides such a quality or power* that when its particles come into contact with the organ of smell, in the absence of preventing causes, the sensation of smell takes place.

c. Next of the sense of Taste (*rasana*).

Aphorism II.

The organ of Taste resides in the tongue.

a. Particles of sapid bodies taken into the mouth, and dissolved in a fluid such as the saliva, come into contact with the organ, and then, in the absence of preventing causes, the sensation of taste takes place.

b. So, that body in which resides the power of producing such a sensation is called sapid.

c. Next of the sense of Sight (*chakshush.*)

Aphorism III.

The organ of Sight resides in the eye.

a. Light comes to the eye from bodies either self-luminous or illuminated, and then, in the absence of preventing causes, the sensation of sight takes place.†

---

* This assertion is designed as a demurrer to the dogma of the Nyáya that Odour belongs only to an Element called Earth (*prithiví*), and that everything that is odorous is therefore earthy.

† Here we oppose the Nyáya theory that a ray *goes* from the eye to the object.

b. Of the nature of Light, and its relation to Colour, and of the structure of the Eye, we shall have occasion to treat in the sequel.

Next of the sense of Touch (*twach*).

Aphorism IV.

The sense of Touch, by which the tangible is apprehended, distributed throughout the whole body, is particularly delicate at the tips of the fingers ; for it is chiefly in the fingers that there resides the power of ascertaining, by touch, the number and magnitude [of small objects*].

a. Next of the sense of Hearing (*s'rotra*).

Aphorism V.

The sense of Hearing is lodged in the ear.

a. What is it that is apprehended by this sense ? To this we reply ;—that the sensation of *sound* takes place when a membrane in the ear [—the tympanum—] vibrates in a manner adapted to the sense of Hearing.† Of the nature of sound, and of the structure of the ear, we shall have occasion to treat in the sequel.

---

* For example,—if the points of a pair of compasses are separated to about the tenth of an inch, their separateness and their distance can be distinctly appreciated by applying them to the tip of the finger, but not by applying them to the cheek.

† Thus we demur to the theory that Sound is a *Quality* residing in an all-pervading Element called the Ether, and that it may be therefore at all times in a sonorous body, whether perceived or not.

SECTION II.

OF THE OBJECTS OF THE SENSES.

c. Now we shall make some remarks on the *Objects* of the Senses, the first of which, in the enumeration given in Aph. XIV. Book I., is Odour (*gandha*).

Aphorism VI.

The quality of Odour is that power belonging to certain substances whereby the sensation of Smell arises when the particles of such a substance come into contact with the Organ of Smell.

a. As there is no proof that whatever possesses Odour must be of one particular kind of substance—viz. of Earth—, and as there appears to be no advantage in holding such an opinion as a hypothesis, we shall simply speak of anything as being odorous from which we derive the sensation of Smell, whether it be solid or fluid. On this point we shall have occasion to speak further in the sequel*.

b. Next of the quality of Savour (*rasa*).

Aphorism VII.

The quality of Savour is that power belonging to certain subtances whereby the sensation of Taste arises when the particles of such a substance come into contact with the Organ of Taste.

a. As pure Water, in our opinion, has no taste, we differ from those who say that it has a sweet taste ; but we hold that Water, in the shape, for example, of the saliva, though itself without taste, conduces to the production of the sensation

---

* When we come to the subject of Chemistry.

by dissolving sapid bodies so as to bring their particles into contact with the Organ of Taste.

b. Next of the quality of Colour (*rúpa*).

Aphorism VIII.

The quality of Colour is the power, belonging to certain kinds of substances, to affect Light in such a manner that when the Light comes in contact with the Organ of Sight the sensation of Colour takes place.

a. What modifications Light is liable to, when it comes into contact with various substances, will be considered in the sequel.*

b. Now, as we talk of seeing a white cow or a green tree, so we talk of seeing a *distant* tree or other object. Do we, then, *see* distance as we see colour ? On this point we proceed to remark :—

Aphorism IX.

The Distance of an object is not seen, but is inferred from the indistinctness of the object and our knowledge of its character, &c.

a. ' Of an object',—i. e. of any perceptible object,—the distance ' is not seen'—i. e. is not directly perceived. How then is it *known ?* To this we reply ' but it is inferred,'—or is learned by inference, from its indistinctness &c. In short the knowledge of distance is due to a general proposition, viz., that in general what is near appears distinct, and what is distant appears indistinct, as is the case with a large tree in the distance, seen through a very small window which occupies a larger space in the field of

---

* When we come to the subject of Optics,

vision than the tree does. Here the tree is not really the smaller of the two, though it *appears* to be so ; and so, from its indistinctness &c. its distance is inferred. The sun is ninety millions of miles distant, but the sun is not indistinct; and therefore, though it is immensely larger than the earth, it appears to
the eye to be at no great distance, and smaller than a potter's
wheel.

b. By the ' &c.' in the aphorism, we refer to the apparent
smallness of an object which we know antecedently to be large.
Ignorance of the cause of the indistinctness of an object gives
rise to mistaken notions of its size and distance. Thus, in a fog,
or when the light is failing, a crow close at hand may be mistaken for a distant elephant, if we do not advert to the cause of
the indistinctness*.

b. Next of the quality of Tangibility (*sparśa*).

Aphorism X.

The quality of Tangibility is the power belonging to certain
kinds of substances to produce the sensation of Touch when they
are brought into contact with the organ of Feel which resides in
the fingers &c.

a. Next of the quality of (*śabda*) Sound (or rather Sonorousness).

Aphorism XI.

The quality of Sonorousness is the power of occasioning such
vibrations as produce the sensation of Hearing when they are
communicated to the tympanum of the ear.

---

* Cf. Reid's collected works, p. 191.

a. The vibrations of a sounding body are communicated to the tympanum by means of some medium, such as the air, as will be explained in the sequel.*

b. Now, as we talk of hearing a harsh sound or a soft sound, so we talk of hearing a *distant* sound. Do we then *hear* the distance as we hear the sound ? On this point we proceed to remark :—

Aphorism XII.

The Distance of a sounding body is not heard, but is inferred from the character of the sound and our knowledge of the sounding body.

a. A sound, to one near the sounding body, appears louder than a like sound to the same person when at a greater distance. The sound of a waterfall in the Himálayas, while it continues the same to a person so long as he remains standing near it, becomes less and less as he moves away from it; so that, recollecting the force of the sound when he stood near it, he can form some estimate of its distance, when he perceives the force of the sound to be more or less diminished. So, on the other hand, when we know that the cause of a sound is very distant, we infer that the sound is much greater than it appears. For example, when we know that a certain sound has arrived from a distance of ten miles, we probably infer that it is the sound of a cannon, though it may not appear so loud as that of a musket fired near us. Hence, again, when we know neither the distance nor the nature of the cause of the sound, we may confound the thunder of a distant cloud with the rumbling of a neighbouring cart.

b. Now the substances which affect the senses by means of

---

* When we come to the subject of Acoustics.

these five powers or qualities, may be regarded in two ways, as
we proceed to explain.

SECTION III.

OF MATTER, OR THAT THE QUALITIES OF WHICH FUR-
NISH THE OBJECTS OF THE SENSES.

Aphorism XIII.

Substances which affect the senses (i. e. material substances,*
or matter,) may be regarded as being either in the shape of Atoms
or of Masses.

a. We shall consider Atoms in the sequel.† In the meantime we proceed to divide Matter as it occurs in mass.

---

* Of such substances the Nyáya reckons five—its five Elements (*bhúta*).
As the term ' Element' is not equivalent to the term ' Matter',—not being applicable to composite material masses such as trees, human bodies, &c. ; and
as there is no other term common to the five which is not at the same time
common to things *not* material, it follows that the philosophical terminology of
the Nyáya offers no term co-extensive with the term Matter. No practical
inconvenience, however, need result from this, if the fact be carefully kept in
mind—along with the exact reasons why the fact is so. That this is the case it
will be our business to endeavour to show as we proceed. ' Material things',
in the present section, are rendered *bhautika-padártha* i. e. ' things formed of
the Elements.'

† When we come to the question of the Constitution of the Masses.

Aphorism XIV.

Matter viewed in mass, may be divided into this Earth with its inhabitants, and the rest of the material universe.

a.   In accordance with this division we may define the province (or the object-matter) of Astronomy as follows.

Aphorism XV.

The positions and motions &c. of the rest of the material universe (—the sun &c.—), viewed in connection with the Earth regarded as a unit, constitute the object-matter of Astronomy.

a.   As the reader may be curious to learn the opinions of Europe on the subject of Astronomy, we shall give an outline of these before proceeding further with the division of the province of the sciences.

SECTION IV.

OF THE FACTS OF ASTRONOMY.

b.   The first point here to be attended to is a two-fold division of the heavenly bodies, which we may state as follows :—

Aphorism XVI.

These may be divided into two sets,—for there are some that shine by their own light, and some that shine by a borrowed light.

a. That is to say—these heavenly bodies are of two kinds:—
why ?—because some of them, as the fixed stars, shine by their
own light as the sun does ; while others, as the planets, shine by
a borrowed light, as the moon does by that of the sun.

b. If it be asked—why, then, do not the planets change their
aspect as the moon does ?—we reply that they *do* change their
aspect, as the moon does, to one so placed as to observe the
change ; and, although this change is not discernible by the naked eye, yet by the aid of the telescope it is easily rendered evident in the case especially of the planet Venus.*

c. Next we have to declare the figure of the Earth.

Aphorism XVII.

The Earth is a globe.

a. One proof of this is exhibited in the chapter on Demonstrative Exposition in the first Book.   Other proofs will be referred
to in the sequel.   That the Earth is a globe, has been shown by
BHÁSKARA ÁCHÁRYA in his *Goládhyáya.*

b. Now, as regards the motion of the Earth :—

Aphorism XVIII.

The Earth's rotation on its axis produces the alternation of
day and night, whilst its annual revolution round the sun produces the change of the seasons.

a. Why then does the Sun appear to move round the Earth ?
The Sun appears to move round the Earth just as, to children in

---

* The pandits of Benares have had an opportunity of seeing this by means
of the College telescope taken into the city for the purpose.

a boat which is receding *from the bank,* the bank appears to recede *from the boat.* So ÁRYYA BHAṬṬA says, " As he who stands in a moving ship sees the mountain approaching him in the opposite direction, so does the inhabitant of Ceylon see the fixed stars wending toward the West." Proofs of the fact that the Earth moves round the Sun, will be given in the sequel.

b. Now, as regards the motion of the Moon.

Aphorism XIX.

The Moon moves round the Earth.

a. This is evident. Now, how and when do eclipses take place ? In regard to this we have to declare as follows :—

Aphorism XX.

When the moon, revolving round the earth, comes between the Sun and the Earth, then there is an eclipse of the Sun ; and when the Earth comes between the Sun and Moon, then there is an eclipse of the Moon.

a. As the Moon moves round the Earth once in a month, it might be expected that eclipses should occur once in a month. Why this does not happen will be explained in the sequel.

b. Now, as regards the motions of the Planets.

Aphorism XXI.

The Planets move round the Sun.

a. Why, then, it may be asked, do they not appear to do so ? One reason is because the Earth also moves round the Sun ; and therefore the Planets, as viewed from the Earth, do not appear to move in such paths as it is ascertained by calculation that they really do move in.

b. As the earth is attended by a Moon, it might be expected that the other Planets also should be attended by moons. In regard to this, we have to state that the planets which are nearer the Sun than the Earth is, have no moons, but some of the more distant planets are furnished with several. With regard to the number of moons attending these we have to remark as follows :—

Aphorism XXII.

The planet Jupiter has four moons ; and the planet Saturn, besides seven moons, is attended by a large luminous ring surrounding him.

a. Saturn's ring is not clearly discernible without a telescope.

b. By the telescope, myriads of stars, in addition to those visible to the naked eye, are rendered visible. Thus the nature of the Milky Way has been discovered, in regard to which we have to remark as follows :—

Aphorism XXIII.

The Milky Way is nothing else than a conglomeration of stars.

a. Other facts regarding Astronomy, not mentioned here or in the sequel, will be found stated in the astronomical treatise of Pandit Bápú Deva.

b. Reverting to the division indicated in the 14th and 15th aphorisms of the second Book, we shall now consider the Earth with its inhabitants not as a partial unit in a system but as a system in itself.

Aphorism XXIV.

The Earth may be considered either as to its surface arrange-

ment, or its constituent masses, or the causes of the existing arrangement ; and this gives rise to a three-fold division of the
subject.

a. We have first, then, to give some account of the surface
arrangement of the Earth.

SECTION V.

OUTLINES OF GEOGRAPHY.

Aphorism XXV.

The arrangement of the Earth's surface furnishes the object-
matter of the science of Geography.*

a. The surface of the Earth is made up of sea and land,
with mountains, rivers, cities, &c. ; and the science that is concerned about the relative situations of these is Geography.

b. Now, rivers do not accomodate their courses to the positions of cities, but large cities are generally built, for the sake
of water, on the banks of rivers,—as the city of Benares on the
bank of the Ganges. Mountains, again, are not distributed according to the course of rivers, but the rivers take their course
in accordance with the distribution of the mountains.   With the
view of explaining how this happens, we first remark as follows :—

---

* The science of the Earth's surface—*bhú-prishtha-vidyá.*

Aphorism XXVI.

The flowing of rivers is dependent upon mountains, and it is by the arrangement of the mountains that the courses of the rivers are determined.

a. Under the term ' mountains' we here include all high grounds. Great rivers like the Ganges, take their rise in great mountain-chains, like the Himálaya. In explanation of the fact that great rivers take their rise in great mountain chains, we have to state what follows.

Aphorism XXVII.

The tops of lofty mountains are cold, and there is a limit beyond which if a mountain rises it is always covered with snow from that limit to the top.

a. This limit, which is called the ' snow-line' or the ' limit of the snows' (*hima-símá*), differs in height, in different countries, according to the heat of the climate. In the Himálaya mountains (on the southern face) it is about 16,000 feet above the level of the sea. In the south of Europe, where the climate is colder than that of India, the line of perpetual snow is only about 1600 feet above the level of the sea ; while, near the poles, snow covers the land down to the water's edge.

b. A natural consequence of a river's taking its rise in a snowy range next requires our notice.

Aphorism XXVIII.

Even when there is no fall of rain, those rivers which arise from the melting of the snow of lofty mountains become swollen in the hot weather.

a. This is observable in the Ganges at Benares, and the reason of the increase of the water must be evident to those who reflect that the river comes from the melting of the snows of the Himálaya mountains.

b. As the Ganges is flooded during the summer, so too is the river Nile in Egypt. The following remarks relative to the cause of the flooding of the Nile are taken from the work of the Grecian historian Herodotus, who lived 427 years before Vikramáditya (B. C. 484).

c. " Concerning the nature of this river," says Herodotus, " I was not able to learn anything, from the priests or from any " one besides, though I questioned them very pressingly. For " the Nile is flooded for a hundred days, beginning with the " summer solstice; and after this time it diminishes, and is, " during the whole winter, very small. And on this head I was " not able to obtain anything satisfactory from any one of the " Egyptians, when I asked what is the power by which the Nile " is in its nature the reverse of other rivers. Yet there are " Greeks who, wishing to appear very wise, have offered three " explanations of the peculiarities of this river. Of these expla- " nations one is as follows :—That the north-wind, which blows " in the summer, is the cause of the rise of the river by pre- " venting it from discharging itself into the sea. But often it " has happened that the north-wind has not blown ; yet the " Nile has risen as high as ever. Besides, if the north-wind were " the cause, it would follow that all rivers which flow against the " north-wind must exhibit the same effect ; and so much the " more as their streams are feebler. But there are many rivers " exposed to this ,wind which undergo no such change as that " which takes place in the Nile.

d. " The second explanation is this that the rise of the Nile " happens because the Nile flows from the ocean, which, as they

G

" say, encompasses the whole earth.   But as to what is here said
" of the ocean, it is an obscure fable, destitute of proof.   I know
" of no such river as the Ocean.   Homer, perhaps, or some of
" the earlier poets, finding the name, transported it into the
" language of poetry.

e.   " The third explanation is this, that the overflow of the
" Nile arises from the melting of snows.   But now, how can it be
" that a river which rises in Lybia, passes through Ethiopia, and
" discharges itself in Egypt—thus proceeding from the hotter to
" the cooler regions,—should owe its rise to snows ?   There are
" many reasons which may convince any man that this cannot be
" the case.   In the first place—and it is a sufficient evidence to
" the contrary—the wind that blows from these regions is hot.
" Again, the men of those countries are blackened with the
" heat.   Besides, kites and swallows remain there through the
" year, while cranes, flying from the Scythian winter, take up
" their abode there during that season.   But, of necessity, none
" of these things would happen if, in the countries through
" which the Nile runs, and where it takes its rise, snow fell even
" in the smallest quantity."

f.   From all this it is clear that Herodotus was not aware of the
difference between the climate of high mountains and of the
plains in a torrid region, which is declared in the 27th aphorism
of this Second Book.   Those, on the other hand, who are acquainted with this fact in regard to mountains, would at once infer,
from the overflow of the Nile in summer, that the river takes its
rise in a snowy range ;   and that it does so, just as the Ganges
does, has now been ascertained by inspection.

g.   Now, he who travels among the Himálaya mountains, and
sees, one after another, innumerable peaks and innumerable
streams descending from them, is apt to think that there is no

order or arrangement in the matter :—but if he takes a comprehensive view of the whole at once, he will see that there is a certain amount of uniformity in the arrangement throughout, the nature of which we now proceed to declare.

Aphorism XXIX.

The Himálaya range of mountains extends, East and West, along the North of India.

a. In this range there are various roads by which travellers can pass the mountains into the countries beyond. The line of snowy mountains in which these passes occur is therefore called " The Ghát-line of the snows"—[the word *ghát* meaning a pass].

b. Next, of the great peaks.

Aphorism XXX.

From the Ghát-line of the Himálayas, the highest peaks, with their attendant ridges, extend southwards.

a. Beginning with the peak of Jamnautrí, where the Ganges rises, and reckoning eastward, these southward-pointing ridges are those of Jamnautrí, Nandadeví, Dhawalagiri, Goswámisthána (or Gosa'ín-thán), Kanchangiri, Chamalárí, and the Gemini.

b. Next as regards the distribution of the rivers belonging to these.

Aphorism XXXI.

Between these ridges the numerous streams gradually unite to form a single river.

a. Between the ridge belonging to the peak of Jamnautrí, which is 25,669 feet high, and Nandadeví, which is 25,598, the streams flowing down from the melting snow, gradually converge in consequence of the concave form of the region ; and thus, although there are many considerable hills between these two ridges, such as the peak of Kedárnáth near Gangautrí, all the principal streams unite into one, and form the Ganges.

b. In like manner, between the ridge belonging to the peak of Nandadeví and that belonging to the peak of Dhawalagiri, many streams unite to form the Karnátí, called also the Ghar-gata (or Ghogra), and this falls into the Ganges.

c. Again, between the ridge belonging to Dhawalagiri, and that of Goswámisthána, there are the rivers called collectively the Sapta-gandakí; and these, uniting into one stream which bears the name of the Gandaka, next join the Ganges.

d. Between the ridge of the Goswámisthána and that of Kanchangiri, there are the rivers called collectively the Sapta-Kauśika; and these uniting next join the Ganges in one stream which bears the name of the Kauśí. And so in the other valleys of like formation.

e. Now, on leaving the mountain valley, what kind of ground do these rivers meet with before reaching the Ganges? On this point we have to remark as follows :—

Aphorism XXXII.

These rivers cut their way through the range of sandstone hills that lies parallel to the Himálaya.

a. This range of sandstone hills is not more than from 300 to 600 feet above the ground on either side.

Aphorism XXXIII.

Above the sandstone range are the Dhúns, below is the Bhá-var, and below that the Tará'i.

a. The mean breadth of the Himálaya is about ninety miles. Let this be divided into three portions, of thirty miles each, and it will present three climatic divisions, the lower, middle, and

upper. Let the lower region consist of the Tará'í, the Bhávar
(or Sál forest), and the sandstone range with its Dhúns (or valleys),—i. e., let it occupy from the level of the plains to about
4000 feet above the sea :—the middle region will consist of the
ground from there to about 10,000 feet above the sea; and the
upper region, of the ground above that.

b. Now, we mentioned before (*Aph.* 27) that the tops of
high mountains are cold; and it is evident that if there be snow
on the top of a mountain, and the air be warm at the foot, the
temperature must be gradually colder from the foot to the top.
Now some animals love heat, and others love cold; and some
plants love heat, and others love cold; and both animals and
plants thrive best in those places where they find the temperature which they prefer. This is found to be the case all over the
world ; and how it is exemplified in the Himálaya we proceed
to state :—

Aphorism XXXIV.

Plants and animals are different at different heights on the
Himálaya, as is the case with other high mountains also.

a. Thus, as regards vegetation,—in the lower region there is
the Sál (Shorea), the Síssú (Dalbergia), the Tund (Cedrela), the
Paláśa (Butea), the Banian (Ficus Indicus), the Peepul (Ficus
religiosus), &c. In the middle region, instead of these, there
are trees* that grow in such climates as that of England, and
which will not thrive in warm climates. In the upper region
there are pines, birches, and the congeners of such other plants† as

---

* Oaks, Alders, Cherries, Pears, &c., correct names for which are not has-
tily to be determined in the plains.

† Junipers, larches, &c.

love a degree of cold even greater than that of England.

b. As regards zoology,—in the lower region live such animals;—men* &c,—as love heat. In the middle region live such† as suffer in the heat of the lower region. In the upper region live such‡ as cannot thrive in the heat even of the middle region.

c. Then again, in the lower region there are elephants, rhinoceroses, tigers, and deer. In the middle region there are no elephants, rhinoceroses, or tigers, and only one kind of deer. In the upper region there are none of these elephants, deer &c., while in that region alone of the three are found the wild goat and the wild sheep. Then again, there are in the upper region no such crows as are seen in the lower one, and very few of these are met with even in the central one.

d. Thus, within the limits of the Himálaya, in consequence of the increasing coldness of the climate from the base upwards, there are to be found such climatic differences as are not usually to be met with except at wide intervals on the earth's surface.

e. We shall now mention the great divisions of the earth's surface as recognised by Europeans.

Aphorism XXXV.

The dry land is divided into four portions, called respectively Asia, Africa, Europe, and America.

---

* The Koch, Bodo, Dhímal, &c.

† The Lepchas &c.

‡ The Bhotiyas.

a. The earth is in form, nearly a sphere, with a diameter of about eight thousand miles. More than two-thirds of the surface of the earth is covered by the ocean. Of the remaining dry land the great divisions are named as just mentioned.

Now, how are these four to be recognised?

Aphorism XXXVI.

Asia is the division in which India is situated, and Europe that in which is England:—Africa and Europe lie to the west of Asia, Europe being to the north of Africa:—America extends from north to south over great part of the hemisphere that is antipodal to India.

a. The forms and the relative situations of the various countries and seas on the globe are represented in maps, from which they may be learnt.

b. Some countries are warm, and some are cold. With regard to the principal reason of this. we have to observe as follows :—

Aphorism XXXVII.

The climate of a country depends mainly upon its receiving the sun's rays perpendicularly or obliquely.

a. In countries near the equator, the sun's rays fall nearly perpendicularly upon the earth ; and these countries are warm, as India. Countries placed nearer the pole receive the sun's rays obliquely, hence in smaller quantity, and are comparatively cold, as England.

b. We have already explained how rivers take their rise in mountains, as the Ganges &c., in the Himálaya mountains.

Now, in Africa the mountains are few; hence the rivers are few, and much of that region is consequently a desert.

c. In books on Geography, in addition to the relative positions of towns and countries, the religion, habits, forms of government, &c., of the inhabitants are detailed. For fear of prolixity we shall not here go into these topics.

d. Having now given some account of the surface-arrangement of the earth, we have next (—as intimated in APH. 24. BOOK II.—) to consider its constituent masses. Masses (—as hinted in *Aph.* 13. BOOK II.—) we regard as the products of Atoms, to which belong three important properties, as we proceed to indicate in the aphorism here following.

SECTION VI.

OF MATTER AND MOTION.

Aphorism XXXVIII.

Now the enquiry has reference to Atoms, and their Attraction, Repulsion, and Inertia.

a. *Atoms?* Every material mass is divisible into very minute particles, indestructible by human power. For example, a piece of iron, or other metal, may be bruised, broken, cut, dissolved, or otherwised transformed, a thousand times; but can always be exhibited again as perfect as at first. Let the most minute resisting particles of which the mass is composed be called Atoms —i. e. what cannot be further cut or divided.

b. *Their Attraction?* It is found that the Atoms, whether separate or already joined into masses, tend towards all other

Atoms or masses,—as when the atoms of any mass, such as a stone, are held together (by some influence which we know only by its effects) with a certain force; or when a block of stone is similarly held down to the earth on which it lies; or when the sea (in the flow of the tide) rises towards the moon. Let the cause of these effects be called *Attraction.*

c. Repulsion? Atoms, under certain circumstances,—for example when Heat is diffused among them,—have their mutual attraction resisted, and they tend to separate,—as when ice heated melts into water, or when water heated expands into steam. Let the cause of these effects be called *Repulsion.*

d. Inertia? As a potter's wheel, when made to revolve, at first offers resistance to the force moving it, but gradually acquires speed proportioned to that force, and then resists being again stopped—in proportion to its speed, so all bodies and atoms appear to have, in regard to motion or rest, a *stubbornness*, tending to keep them in their existing state whether of motion or of rest. Let the cause of this be called *Inertia.*

e. Now we have to see how the enquiry into Atoms and their mutual Attraction, &c., is to throw light on the nature of the material universe; and we proceed in the first place to declare as follows.

Aphorism XXXIX.

The sensible universe is formed of Atoms.

a. As before mentioned, a piece of metal, after having been bruised and broken and dissolved and altered in a thousand ways, can be always recovered as perfect as at first. Although this is not the case with things that are organized,—as the leaves of trees or the feathers of birds,—for we cannot restore to an organized

H

structure the form that it had before,—yet, as we shall see when
we come to the subject of Chemistry, even when these are burn-
ed, not a single Atom is really lost.

b. Now, with regard to the size of the Atoms.

Aphorism XL.

Atoms are extremely minute.

a. Among the proofs of this fact it may be mentioned that
goldbeaters, by hammering, reduce gold to leaves so thin that
360,000 must be laid upon one another to produce the thickness
of an inch. The thickness of even one of these leaves is, we
know, much greater than that of an Atom, because gold can be
spread out much thinner upon silver wire ; and even then there
is no proof that the thickness is no more than that of an Atom.

b. The followers of the Nyáya hold that the magnitude of an
Atom is *smaller than any magnitude* (or *positively small*), so that
the magnitude of an Atom is no element in the magnitude of a
combination of two Atoms, because that which is produced from
(the intensifying of) the *small* ought to be still smaller (—as in the
multiplication of fractions by fractions,—or in the summation
—as regards the positive result—of *minus* quantities in Algebra).
We on the other hand (for the most part) hold, that, although
exceedingly small in comparison with the masses formed of them,
yet there is no reason why they should not possess a determinate
bulk with reference to these,—so that the bulk of two Atoms
may be twice the bulk of one. Reasons for holding this opinion
will be mentioned in the sequel.*

c. A European philosopher, of the name of Boscovich, put for-

---

* When treating of crystallization.

ward a theory that Atoms have no magnitude, and that they [differ from mathematical points only in as much as they] have primary force* [—or a repulsion which secures to each atom its own range]. Although this hypothesis is not devoid of plausibility, yet it does not serve so well to explain the varied phenomena of the material universe as the hypothesis followed in the present work :—we therefore proceed to lay down the following position.

Aphorism XLI.

Matter always occupies space.

a. A portion of matter, as a large mass of cotton, by being compressed, can be made to occupy a smaller space, but it cannot be so compressed as to occupy *no* space. In the same way all other bodies on no occasion altogether part with magnitude. This property of never ceasing to possess magnitude is called the *Impenetrability* of matter,—a term implying that no two or more bodies can occupy exactly the same place at the same time. The following are examples of this.

b. If a stone be dropped into a jar brimful of water, then water will flow out to make room for the stone.

c. A glass tube, left open at the bottom, while the finger closes the top, if pressed from air into water, is not filled with water, because the air contained in it resists; but if the air be allowed to escape by removing the finger from the top, then the tube becomes filled to the level of the water around it.

d. Invert in a vessel of water a smaller vessel full of air. Some water will enter; but the water cannot fill the smaller ves-

---

* See Preliminary Dissertations to the Enc. Brit. p. 606.

sel until, by turning its mouth upwards, we allow the air to escape. If the vessel of air be inverted over a floating lighted taper, this will continue to float under it, and to burn for some time in the contained air, however deep in the water it be carried. In like manner, in a large inverted vessel of air, suspended by a rope, men can be lowered to the bottom of the sea to recover sunken property. The vessel for this purpose, being usually shaped like a bell, is called a Diving-bell.

e. Now there must be some reason why Atoms do not remain separate, and therefore we conclude as follows.

Aphorism XLII.

Atoms attract each other.

a. In other words, there is a mutual *Attraction* among Atoms. Attraction exists also between the separate masses formed of atoms. Of this fact the following are instances.

b. Logs of wood floating in a pond, or ships in calm water, approach each other, and afterwards remain in contact.

c. As stones fall, and plum nets hang, in India, towards the centre of the earth, so is it on the side of the globe opposite to India also. People there, therefore, like flies on the opposite side of a pane of glass, are standing with their feet towards us. Bháskara Áchárya, in his *Goládhyáya,* illustrates this as follows.

"This (globe of the earth) is covered on every part with multitudes of mountains, groves, towns, and monuments, as is the ball of the Nauclea's globular flower with its multitudes of anthers."

d. "But then," some one might object, "men are not fastened into the ground as the anthers of the Nauclea flower are fastened into their globular receptacle so that they do not fall

away from it. How, then, do the men not fall off who are standing at the distance of a fourth part of the circumference from
us, or in the opposite hemisphere ?" To this Bháskara Áchárya
replies as follows.

" Each one, wherever he stands, regards the earth as being
under him, and himself standing *upon* it :—therefore those
that stand at the distance of a fourth part of the circumference
from one another mutually regard one another as standing *hori-*
zontally, and those that stand in the opposite hemisphere as having their heads downmost (and forming antipodes) like the man
and his reflected image at the margin of a piece of water;—and
thus those who are (as regards our own position) placed horizontally or head downmost, stand *there* just as we do *here.*"

e. It is in consequence of all the parts tending to a common
centre that the earth is spherical.   In like manner when particles
of mist coalesce into drops, these are spherical.   They lose their
spherical form, of course, when they reach the ground ; but melted lead, allowed to rain down from an elevated sieve, by cooling as
it descends, retains the form of its liquid drops.   It is in this way
that the spherical small-shot used by sportsmen is made.   Lead,
when allowed to drop from a height, does not assume a spherical
form if it is not in a fluid state.   Hence we are led to conjecture
that the earth, the moon, the sun, and the planets, all of which
are round, must at some time have been to a certain degree fluid ;
and also that they are all subject to the same law of Mutual
Attraction.

f. But it may be asked,—if all things on this globe tend to
the centre, why is smoke seen to ascend ? Smoke does not, of
its own accord, ascend ; but in reality the air near the earth, being heavier, sinks below it and forces it to rise, just as the water

in a jar causes a straw to rise which has been thrust to the bottom of it.

g. The cause which we call Attraction acts at all distances. The Moon, though 240,000 miles distant from the earth, by her attraction, raises the water of the ocean under her, and thus occasions what is called the *tide.* The sun, still further off, has a similar influence. When the sun and moon act in the same line of direction, (as they do twice in each month,) the tide is greater (—and is then called a spring-tide).

h. But, although Attraction acts at all distances, it does not at all distances act with equal force; and, in regard to this, we proceed to remark as follows.

Aphorism XLIII.

The Attraction is greater the nearer the bodies are to each other; as the light of a lamp is more intense nearer to the lamp than at a distance.

a. Place a wooden board of a foot square at a certain distance from a lamp, and it will just shadow a board of two feet square behind it at double the distance. Now a board with a side of two feet has four times as much surface as a board with a side of one foot, for it is not only twice as long, which would make it double, but twice as broad also, which makes it quadruple, Light, therefore, at double distance from its source, being spread over four times the surface, has only one fourth of the intensity. For a like reason, at thrice the distance, it has only a ninth part, and so on,—the light becoming less intense the more widely it is diffused, just as a bag of rupees, on being equally divided among a greater number of persons, allows a proportionally smaller amount to each. Now light, heat, attraction, sound, and indeed every influence from a central point, is found to decrease in the proportion here illustrated, viz., as the surface of squares, the one of which shadows the other, increases. Accordingly, what weighs 1000 pounds at the sea-shore, weighs five pounds less at the top of a mountain of a certain height. This is proved experimentally by using a spring-balance. The spring, drawn out to a greater length, indicates a greater weight at the level of the sea than on a lofty mountain. By calculating, according to the proportion of weight lost on a hill of a certain height, it is found that the weight, or tendency towards the earth, of 1000 pounds, must be diminished to five ounces at the distance of the Moon. This fact will be reverted to in the sequel.

b. In pouring water from a jar, it is seen that the water, in consequence of the attraction of the jar, does not readily leave it, but is inclined to run down the outside of the jar, which attracts it. It is in order to remove the water from the attraction of the vessel, that various vessels are furnished with spouts,—as is the case with the water-pots of Hindú ascetics. When a vessel has not a spout, you may pour from it without spilling, by leaning the lip of the vessel against a rod of glass, or the like, down which the liquid will readily run,—the attraction of the rod then counteracting that of the vessel.

c. The particles of water cohere among themselves in a degree which causes small needles, gently laid upon the surface, to float. The weight of the needles is not sufficient to overcome the cohesion of the water-surface. For the same reason many light insects can walk upon the surface of water without being wetted.

d. Between two plates of glass, or the like, standing near to each other with their lower edges in water, the water, attracted by them the more strongly in proportion to their closeness, rises

above the level of that on the outside.   In the same way, water,
ink, or oil, coming in contact with the edge of a book, is rapidly absorbed far inwards among the leaves; and the wick of a
lamp lifts the oil, to supply the flame, from two or three inches
below it.

    e.   It has been already remarked that when heat is diffused
among the Atoms, they tend to separate.   On this point we proceed to declare as follows.

Aphorism XLIV.

    The Atoms are more or less close, according to the quantity
of heat, producing repulsion, among them ; and hence the forms
of solid, liquid, and air.

    a.   But what is Heat?   This we are not at present prepared
to say.   We shall consider Heat at present merely as a cause of
Repulsion.

    b.   When a continued addition of heat is made to any body,
it gradually increases the mutual distance of the constituent
atoms, or dilates the body.   A solid thus is at first enlarged and
softened ; then melted or fused, that is to say, reduced to the
state of liquid, as the cohesive attraction is overcome ; and
lastly the atoms are repelled to still greater distances, so that
the substance is converted into elastic fluid or air.   Abstraction
of heat from such air causes return of states in the reverse order.

    c.   Thus ice when heated becomes water, and water when fur-
ther heated becomes steam.   The steam when cooled becomes
water, as before, and the water when cooled becomes ice.   Ice,
water, and steam, therefore, are three forms or states of the
same substance.   Other substances are similarly affected by heat ;
but as some require more heat to liquify them, and some less, it

is thus that we find the existing variety of solids, liquids, and airs.

d. We have now to consider some peculiarities of state which depend on certain modifications of Attraction and Repulsion.

Aphorism XLV.

Certain modifications of Attraction and Repulsion produce the peculiarities of state called Crystal, Porous, Dense, Hard, Elastic, Brittle, Malleable, Ductile, and Tenacious.

a. Crystallization? When various kinds of salts are dissolved in water, and the water is then allowed slowly to evaporate, each salt assumes a peculiar form and no other. From finding that certain substances thus invariably assume a regular form, we infer that, at the time when a mass is beginning to be formed, the Atoms do not attract one another equally all round, but that the Attraction acts between certain sides or portions, thus leading to the regular arrangement which we call the crystalline. The name is taken from rock-crystal, which is found in strikingly regular forms.

b. Porosity? Even among the particles of hard stones there are vacant spaces. Hence a kind of sandstone, suitably shaped, serves to filter water. Let the cause of its being fitted for this be called its Porosity.

c. Density? A cubic inch of mercury is nearly fourteen times heavier than a cubic inch of water. It is inferred that a greater quantity of Atoms exists in a given bulk of that which is the heavier. This is expressed by saying that the substance is more dense. When a body dilates or contracts by the addition or abstraction of heat, its entire weight does not change, because the quantity of its atoms does not change; but the weight of a

I

given bulk of it, as a cubic inch, changes, because the quantity of atoms in the cubic inch changes.

d. The comparative *weights of equal bulks* of different substances are called their specific gravities. Water is the standard employed for determining this, as will be shown in the sequel.

e. *Hardness* is that quality which may be measured by the power of one body to scratch another. The diamond is the hardest of known substances. It cuts or scratches every other body.

f. *Elasticity* is that which restores to its former position what had been forcibly altered as by bending.

g. *Brittleness* is the property of a body which, though hard, is easily broken. For example, glass scratches iron, so that it is harder than iron; yet it is very easily broken.

h. *Malleability* is the property, belonging to gold &c., of being reducible into thin leaves by hammering.

i. *Ductility* means the susceptibility, belonging to gold &c., of being drawn into wire.

j. *Tenacity* means that modification of the force of cohesion among the atoms of any mass, by which it resists being drawn asunder. A great weight can be supported by a thin iron wire; but only a comparatively small weight can be supported by a leaden wire of the same size. Tenacity, differing in degree in different substances, is most remarkably characteristic of iron.

k. As the bodies formed of atoms may be at rest or in motion, we have now to consider the phenomena of motion.

SECTION VII.

OF MOTION AND REST.

Aphorism XLVI.

Motion is the term applied to the phenomenon of the changing of place among bodies.

a. Motion is *straight,* in the apparent path of a falling body,— *curvilinear,* in the track of a body thrown obliquely,—*accelerated,* in a stone falling to the earth,—*retarded,* in the stone thrown upwards while rising to the point where it stops before descending.

b. Now, how is motion produced, and how does it cease?

Aphorism XLVII.

Force is required whether to give motion or to take it away.

a. It has been already mentioned that a potter's wheel first offers resistance to the force moving it, and then resists being stopped. So a person standing upright in a carriage falls backwards when the carriage suddenly moves forward; and the same person standing upright whilst the carriage is moving rapidly forward, falls forwards when the carriage suddenly stops. So too, when a boat crossing a river strikes against the bank, all the ignorant people who in their haste to get ashore have stood up, fall forward towards the bank.

b. But, whether is the state of rest or that of motion natural to bodies? To this we reply :—

Aphorism XLVIII.

Uniform straight motion is as naturally permanent as rest.

a. That is to say, a body can neither acquire motion, nor lose motion, nor bend its motion, without a cause.

b. When an iron ball is rolled on the ground, if the ground be not smooth, the motion is speedily destroyed. If the ground be smooth, then the motion, being less impeded, continues longer; yet it is gradually destroyed by obstacles which cannot here be entirely removed. It is only in the sky that we observe an entire absence of obstructions, and there the motions of the planets go on uninterruptedly.

c. An arrow discharged horizontally at a mark gradually bends downwards to the earth by which it is attracted; but, when the air is still, it does not tend to change its course either to the right hand or to the left:—were it not for this, no one could calculate upon taking a correct aim.

d. A stone from a whirling sling, the moment it is set at liberty, darts off as straightly as an arrow from the bow-string; and it is only because of the great difficulty, even after long practice, of determining the point of the circle from which it should depart, that it is so difficult to hit a mark with it.

e. From such facts it appears that a body moving in a circle is constrained by some force which is contrary to its Inertia. For example:—if a man is seen in the distance moving his hand, and a ball is seen whirling round it, then, though, from the distance or some other reason, we may see no string between his hand and the ball, yet we conclude with certainty that there is a string of some sort.   In like manner, in the case of the circular motion of the moon and the planets, we infer that there is some force interfering with the straight motion which would else result from their Inertia, since a motion other than straight cannot be the result of any single force.

f. We have now, then, to consider the results of more than one force acting upon a body at once. It has been remarked that motion caused by a single force is always in a straight line. It is clear that an additional force applied in the same direction causes a body to move more rapidly in the same direction, as when a boat floating with the current of the Ganges is urged in the same direction by the wind also. When a force is directly opposed by an equal force, motion does not take place; so two bulls of equal strength, opposing their heads with equal force, both maintain their ground. Whichever of two opposing forces is the greater, motion takes place in the direction of that force, with a rapidity resulting from the difference of the two forces; as when a boat, with a fair wind, sails up against the current of the Ganges. These facts are quite obvious. The following is perhaps less so.

Aphorism XLIX.

If two forces act, upon a body, neither in the same direction nor in an exactly opposite direction, the body will move in an intermediate direction.

a. Thus a boat impelled by oars across a river, and at the same time acted upon by the current, moves neither straight across nor straight down the river, but slantingly. Therefore the ferryman, in order to bring his boat to the desired landing-place on the other side of a rapid river, calculating his own strength and that of the current, starts from such a point up the stream that the landing-place may lie rightly between the direction of the current and the direction that he labours in.

Aphorism L.

If two forces act upon a body, by one of which it is projected in a straight line, whilst by the other it is continually directed towards a fixed point, the result is Circular Motion.

a. Thus when a sling, with a stone, is put in motion, the stone, which constantly seeks to move away, moves in a circle because it is restrained towards the hand. If, in the rapid motion, the string break, or if it be let go by the hand, then, since only one of the forces continues to act, the stone will fly off as if it had been thrown straight from the hand.

b. So also in an oil-mill, the ox, when driven, seeking to go forward, and prevented by the beam to which its neck is yoked, moves in a circle round the mill. If he were quit of the yoke, he would probably move off from his driver in a straight line,— his attempt to do this having conduced in the first instance to the circular motion which he is confined to.

Aphorism LI.

The force which pulls a body towards a centre is called the Centripetal Force. The force in virtue of which it tends to recede from a centre is called the Centrifugal Force. Together these are called Central Forces.

a. Phenomena exemplifying the centrifugal force are such as the following.

b. Bodies laid on a whirling horizontal wheel, such as that used by potters, are readily thrown off.

c. In a handmill the grain, carried round by the stone, travels outwards till it escapes as flour at the circumference.

d. When a water-pot, filled with water, is whirled rapidly round by means of a string fastened round its neck, the centrifugal force prevents the water from falling out even whilst the vessel is upside down.

e. A man or a horse, suddenly changing his direction in running (as when turning a corner at speed), instinctively leans inwards, or towards the new direction, to counteract the centrifugal force that would throw him away from it. A carriage, in rapidly rounding a corner, for want of such instinct, is apt to be overturned.

f. When a mass of very soft clay is placed in the middle of a potter's wheel rapidly revolving, the centrifugal force causes the clay to bulge out. Like this ball of soft clay, the Earth, revolving on its axis, has bulged out at the equator, so that the equatorial diameter exceeds the polar diameter by about twenty miles. In consequence of the existence of this additional mass of matter at the equator, the attraction between the Earth and the Sun and the Moon does not act exactly as it would do if the form of the Earth were a perfect sphere. Hence result the precession of the equinoxes and certain other irregularities to be noticed in the sequel.

g. Motion is said to be *uniform* when equal spaces are moved over in equal times. Motion is not necessarily uniform. In regard to this we have to remark as follows.

Aphorism LII.

Motion may be Retarded or Accelerated.

a. Retarded motion is produced by some force acting continuously on a body in a direction opposed to that which first put it in motion, and thus gradually diminishing its velocity. If you throw a stone vertically upwards, the attraction of the earth will gradually diminish its velocity, till at length the upward motion ceases, and the stone begins to descend. The motion of the stone downwards is the converse of what it was upwards. Here we have Accelerated Motion. For, when the stone, by falling through a certain space, acquired a certain velocity, it would move, in consequence of its Inertia, through another equal space in an equal portion of time, without the application of any additional force. But the force of the Earth's attraction is still acting upon it, and thus the velocity is constantly accelerated.

b. When a boy lets a ball drop from his hand, he can catch it again in the first instant; but after a little delay his hand pursues it in vain.

c. When a *Bela* fruit falls from a lofty branch, the eye can follow it for a time, and mark the gradual acceleration of its descent; but soon, from the increasing rapidity of its fall, it is seen only as a shadowy line.

d. Any fluid falling from a reservoir forms a descending stream of which the bulk diminishes, from above downwards, in the same proportion as the velocity of the particles increases. For example, in pouring molasses, or thick syrup, from a considerable height, the bulky sluggish mass, which first escapes, is reduced, before it reaches the bottom, to a small thread; but the thread is moving proportionately faster, and fills the receiving vessel with surprising rapidity.

e. Some bodies are seen to fall to the earth more rapidly than others; but their doing so is owing to the interference of the air. A sheet of paper falls slowly; but if you roll it into a ball, it will fall rapidly. Gold is very heavy, but it appears to float in the air when beaten into thin leaf.

f. A piece of paper, cut to the size of a rupee, and placed on the top of it, reaches the ground along with the rupee,—for the heavier body prevents the air from impeding the fall of the lighter one.

g. In a vessel from which the air has been removed by an air-pump, a rupee and a feather fall with equal rapidity.

h. A stone thrown straight forward is gradually drawn down towards the earth. If the motion caused by the earth's attraction were a uniform motion, like that caused by the current of a river, the stone, like the boat before referred to, would move in a slanting but straight line. As the downward motion, however, is a continually accelerated motion, the stone must continually change its direction, or, in other wards, it must move in a curve, as we see it do. The form of this curve is more distinctly seen when water is spirted from a syringe.

i. Both accelerated and retarded motion is exemplified in the Pendulum.

Aphorism LIII.

The name of Pendulum is applicable to any body so suspended that it may swing freely backwards and forwards.

a. A common pendulum consists of a ball suspended by a rod from a fixed point.

b. A stone suspended by a string will serve to exemplify several of the phenomena of the pendulum. One of the most important of these is the following.

Aphorism LIV.

The times of the vibrations of a Pendulum are very nearly equal, whether it be moving much or little.

K

a. ' Whether it be moving much or little'—i. e. whether the arc described by it be large or small.   This may be shown by suspending a small weight from a peg by a thread of 39 inches in length.   When the weight is drawn to one side and then set free, it will move backwards and forwards in equal spaces of time,—viz. in a second of time, whether it make a wide oscillation as when first set free, or a gradually smaller one.

b.   It is in consequence of this property that the pendulum is employed to measure time.   A common clock is merely a pendulum, with wheel-work attached to record the number of vibrations, and with a weight or spring having force enough to counteract the retarding effects of friction and the resistance of the air.

Aphorism LV.

The length of a pendulum influences the time of its vibration.

a.   The shorter the string of the pendulum, the more rapid is its oscillation : and the longer the string the slower is the oscillation.

Aphorism LVI.

Action and Reaction are equal and opposite.

a.   It is evident that if no action or movement takes place among bodies but in consequence of either Attraction or Repulsion, there must always be at least two bodies concerned, and each must be attracted or repelled just as much as the other, although one will have less velocity than the other, as it may be itself greater, or fixed to another mass.

b.   Thus, if a man in one boat pull at a rope attached to another, the two boats will approach.   The boat which he pulls will *come,* and the boat in which he stands will *go* towards it.　If they be of equal size and load, they will both move at the same rate; and if they be different, the smaller will move the faster, in whichever of them the man may be.

c. A cannon, when fired, recoils with not less momentum in it than the ball has; but, the momentum in the cannon, being diffused through a greater mass, the velocity is small, and easily checked.

d. When a body in motion strikes upon another body, it meets with resistance.　The resistance of the body at rest, will be equal to the blow struck by the body in motion, and exactly opposed to it.　The operation of this principle is most apparent in the case of elastic bodies.　If a ball be rolled along a smooth floor so as to meet a wall perpendicularly, it will return on the same line.　If the first motion be not in a direction perpendicular to the wall, the reflected motion will diverge exactly as much on the other side of the perpendicular.　Thus, if, whilst standing in the left hand corner of a room, we throw a ball against the centre of the opposite wall, the ball, in returning, will move towards the right hand corner,—the angles on each side of the perpendicular being always equal.　This equality of the angles, or law of reflection, which applies to Heat, &c., also, may be expressed as follows :—

Aphorism LVII.

The angle of reflection is equal to the angle of incidence.

a. Other examples of this will be noticed in the sequel.

b. Having thus shown how all the motions visible among bodies are the effects of nothing else than Attraction and Repulsion acting on the Inertia of Atoms, separate or conjoined, under diversified circumstances, we now proceed to explain, by means of these principles, the peculiarities of rest and motion which depend on the *solid* form of bodies.*

SECTION VIII.

PHENOMENA OF SOLIDS.

c. ' *Solid*' is the term applied to a mass in which the mutual attraction of the atoms is so strong that the mass may be moved about as one body, without the relative positions of the component parts being thereby disturbed. From this results the fact next stated.

Aphorism LVIII.

Force, moving part of a solid, must affect the whole or break off the part.

a. An earthenware vessel may be suspended by a portion of its lip (or by its handle), thus proving that the cohesion of the parts is stronger than the weight of the vessel ; but if an attempt be made to lift the vessel quickly by such a part, the part may rise and leave the vessel behind,—because then the *inertia* is acting together with the weight to destroy the cohesion.

b. If any uniform rod be supported by its middle, like a weighing beam, the two ends will just balance each other. If equal weights be hung at the extremities of the arms, the equi-

---

* This is the division of science commonly called *Mechanics.* In regard to solid bodies, the phenomena of Rest are treated under the head of Statics (—properly Stereo-statics—), and those of Motion under that of Dynamics (properly Stereo-dynamics).

librium will still continue.   The middle point in this case is called the *centre of gravity* ; and that there is such a point in every
mass we proceed to declare.

Aphorism LIX.

In every mass, or system of connected masses, there is a point
around which all the parts balance each other, and this point is
called the *centre of gravity* (or of inertia).

a.   When this point is supported, the body does not fall.
When this point is not supported, the body falls to that side on
which the point is.

b.   Thus the centre of gravity of a bar of wood or iron of
uniform thickness is at the middle point.   Support this point on
your finger and the bar will remain supported.   Move the point
to either side and the bar will fall on that side.

c.   A line drawn perpendicularly to the surface of the earth
from the centre of gravity is called the line of direction.   A body remains standing when the line of direction falls within the
base on which the body is placed ; otherwise it falls.

d.   When people in a boat rise up, the centre of gravity of
the whole, that is to say of the boat with its contents, is raised.
A smaller inclination of the boat to one side will then make the
line of direction fall beyond the base.   When there is any risk
of a boat's upsetting, therefore, the passengers ought to avoid
starting up.

e.   The centre of gravity of a sphere is the centre of the
sphere.   Wherever you place a spherical body of homogeneous
material, on a horizontal plane, the centre of gravity being vertically over the point of contact, the sphere will rest.   On the

other hand, if you place the sphere on a slope, the point of contact not being vertically under the centre of gravity, the body
will roll down.

f. The centre of gravity of a body is not necessarily within
the body itself. In a ring, of iron for instance, the centre of
gravity is some where within the hollow in the midst. Suspend
the ring from two several points, and tie two threads across it in
the direction of the two lines of suspension. The ring will be
balanced if you support the point of intersection of the threads.

g. On these properties of matter, and laws of motion, which
we have described, depends the operation of numerous machines
which have been contrived for the saving of human labour.

h. In looking at a potter's wheel revolving, the truth of the
following proposition will become apparent.

Aphorism LX.

In a solid body moving about an axis, like a wheel or a weighing beam, the different parts have different velocities, according
to their respective distances from the axis or centre.

a. It is clear that a piece of soft clay placed on the rim of
the wheel, passes over, in one revolution, twice the space tra-
versed by another piece placed midway between the centre and
the circumference.

b. With reference to this the following is to be borne in
mind.

Aphorism LXI.

Forces, with different speed, may be made to balance one another if they be connected by some solid medium in such a man-

ner that the speed of the smaller shall be the greater, just in proportion to the greater intensity of the other force.

a. Upon this important truth the whole of Mechanics may be said to hinge. On this depends the operation of the *simple machines,* or *mechanical powers,* as they have been called,—the lever &c., which enable man to adapt any species and speed of power which may be available, to almost any work which he has to accomplish.

b. The simplest of the mechanical powers is the *lever.* A lever is a rigid rod of uniform thickness, made, we may suppose, of wood or metal.

c. If the rod be placed upon a support which is called the *fulcrum,* at its centre, the two arms of the lever will balance each other.

d. If equal weights be hung at the extremities of the arms, the equilibrium will still continue. On this principle the common weighing balance is constructed.

e. If the lever be supported not at the middle, but at such a point that the one arm shall be twice as long as the other, then the centre of gravity will be removed to the side of the longer arm; and a weight suspended from the extremity of the longer arm will balance twice its weight suspended from the extremity of the shorter arm.

f. It is evident from this that if the arms of a weighing balance are unequal in length, an imposition in the weight of goods may be effected.

g. By moving the fulcrum still nearer to one end, a weight at the extremity of the longer arm may be made to balance three times, four times, five times its weight at the other end, and so on indefinitely; and a small addition to the weight at the longer arm will enable it not merely to balance but to raise the larger weight. Thus a single person, by applying his strength at the longer end of a lever, is able to raise a weight which the combined force of several men could not move.

h. It is obvious that the extremity of the longer arm must be moved over a great space, in order to raise the extremity of the shorter arm through a small space; and thus, just as much as is gained in power is lost in time.

i. This applies not only to the lever but to all the other mechanical powers. It does not follow, however, that these are useless; for, since we cannot augment our strength, those instruments are highly useful which enable us to reduce the resistance or weight of any body to the level of our strength, at the expense of a corresponding portion of time.

j. Another great benefit resulting from the use of machinery, as already hinted, is this, that it enables us to take advantage of gratuitous forces, such as a current of wind, a stream of water, or the expansive power of steam. When such forces perform our task, we have only to superintend and regulate their operation.

k. So much for the peculiarities of solid bodies.

l. We have next to consider the peculiar properties of matter in the form of *liquid.*

SECTION IX.

HYDROSTATICS.

m. From their deficiency of cohesion, liquids cannot be maintained in heaps. The wind raises water into waves, but these are immediately afterwards destroyed by gravity. Thus :—

Aphorism LXII.

Liquids always tend to have their surface level.

a. As the particles of fluids gravitate independently, they press against each other in every direction, not only downwards but upwards, and sideways. The pressure of fluids upwards, though it seems in direct opposition to gravity, is a consequence of their pressure downwards.

b. When, for example, water is poured into a vessel with a spout, the water in the spout rises to a level with that in the pot. The particles of water at the bottom of the pot are pressed upon by the particles above them ; they will yield to this pressure if there is any mode of making way for the superior particles, and as they cannot descend, they will change their direction and rise in the spout.

c. For the same reason water may be conveyed to every part of a town, if it be originally brought, by means of pipes, from a height superior to any to which it is to be conveyed. Many of the cities of Europe are supplied with water in this way.

d. When a tube with a very small bore is immersed in a vessel of water, the water rises higher in the tube than the surface of the water in the vessel, being attracted by the sides of the tube,

L

In this way [—see *Aph.* 43. *d.*—] the oil of a lamp rises amid the small tubes formed by the threads of the wick ; and water rises in the very small tubes of trees and other vegetables.

e. We have already had occasion to mention [*Aph.* 45. *d.*] the *specific gravity* of bodies.

Aphorism LXIII.

Water is employed as the standard for determining the compative weight, or specific gravity, of bodies.

a. When we say that a substance is light or heavy, we speak comparatively. Thus chalk is light when compared with iron, and heavy when compared with wood.

b. The comparative weight or specific gravity of a body is ascertained by weighing it first in air and afterwards in water.

c. A piece of gold will displace just as much water as is equal to its own bulk ; so that a cubic inch of water must make way for a cubic inch of gold. The gold will weigh less in water than it did out of it, on account of the upward pressure of the particles of the water. Now supposing that a piece of gold weighed nineteen ounces out of the water, and lost one ounce by being weighed in water, the quantity of water which it displaces must weigh that one ounce : consequently gold must be nineteen times as heavy as water.

d. If the body under trial be of the same weight as the water in which it is immersed, it will be wholly supported by it, as was the water the place of which it occupies.

e. A body placed in water rests when it has displaced a quantity of water equal in weight to the weight of the body ; but if the bulk of the body be less than that of its weight of water, as in the case of a ball of clay, it will sink.

f. If the ball of clay be formed into a hollow jar, then it will displace a quantity of water more than equal to it in weight, and it will float.

g. A boat sinks the deeper in the water the more it is loaded, displacing just as much water as is equal to the whole weight ; and thus iron boats can be made to float. Steamboats on the Ganges are made of iron.

h. In ascertaining the specific gravity of a body lighter than water, it is necessary to attach to it a body of known specific gravity sufficiently heavy to sink it.

i. Water, oil, milk, &c. are fluids which possess very little elasticity. Such fluids are called liquids. Other fluids have a remarkable degree of elasticity. Let us now consider, then, the mechanical properties of *elastic* fluids, or Airs.

SECTION X.

PNEUMATICS.

Aphorism LXIV.

The leading character in which air differs from such a fluid as water is its elasticity.

a. By the elasticity of air we mean its power of increasing or diminishing in bulk, accordingly as it is less or more compressed.

b. In consequence of its elasticity, air presses in all directions.

c. An important property, in respect of which air resembles water, is the following :—

Aphorism LXV.

The air possesses weight.

a, The following are proofs of this fact.

When we suck water through a straw, we do it by expanding the chest; this diminishes the compression of the air within our body, which then no longer counterbalances the pressure of the external air on the surface of the water by which the water is consequently forced up the straw.

c. The weight of a small quantity of air may be ascertained by exhausting the air from a bottle, by means of an air pump, and weighing the bottle thus emptied. Suppose that a bottle, six cubic inches in dimension, when the air is exhausted, weighs two ounces; if the air be then re-admitted and the bottle be re-weighed it will be found heavier by two grains; showing that six cubic inches of air weigh about two grains.

d. In order to ascertain the specific gravity of air, the same bottle may be filled with water, and the weight of six cubic inches of water will be found to be 1515 grains : so that the weight of the water to that of air is about 800 to 1.

e. For certain reasons it is supposed that the atmosphere is about 45 miles high. The weight of the upper portions causes the lower portions to be more dense. At the top of a high mountain the air is so thin that consideralbe difficulty is found in breathing there.

f. The height of a mountain may be estimated by means of an instrument called a *barometer,* or measurer of the weight of the air.

g. A barometer may be made thus. Fill with mercury a glass tube about three feet in length, and open only at one end. Then stopping the open end with the finger, immerse it vertically in a cup containing mercury. Part of the mercury which was in the tube now falls down into the cup, leaving a vacant space in the upper part of the tube, to which the air cannot gain access.—This space is therefore a perfect vacuum; and consequently the mercury in the tube is relieved from the pressure of the atmosphere, whilst that in the cup remains exposed to it. Therefore the pressure of the air on the mercury in the cup supports that in the tube, and the mercury will stand higher in the tube when and where the pressure of the atmosphere is greater.

h. The mercury in the barometer usually stands at the height of about 29½ inches.—When the *barometer* is carried up a hill, or to the top of a high tower, the atmospheric pressure being diminished, the mercury sinks. By means of numerous experiments it has been determined how many hundred or thousand feet of additional elevation correspond to a fall of an inch, or two inches, &c., of mercury in the tube, and thus it is that the instrument can be employed for measuring the height of mountains.

i. The mercury also sinks on the approach of a storm, so that it is of great value for giving warning of an approaching storm to ships at sea.

j. The atmospheric pressure which sustains in a tube a column of 29½ inches of mercury, ought to sustain a proportionately higher column of a fluid the specific gravity of which is less.—The specific gravity of mercury as compared with water is 13½. Multiplying this by 29½, the number of inches of mercury, we get somewhat more than 33 feet in height; and accordingly, on making the experiment, we find that the atmospheric pressure sustains a column of water of above 33 feet in height.

k. The relative specific gravity of two fluids which, like mercury and water, are not disposed to mix with each other, may be determined by the following method. Place a glass tube, curved like the goad of an elephant, with the ends of the limbs uppermost, and pour into it so much of each of the two fluids that their point of meeting shall be the lowest point in the curvature of the tube. Then the heights to which the two fluids rise in the respective limbs will be in the inverse ratio of their relative specific gravity. For example, in such an arrangement, an inch of mercury in the one tube will be found to balance $13\frac{1}{2}$ inches of water in the other; so that we conclude that mercury is $13\frac{1}{2}$ times as heavy as water.

l. When such a curved tube as we have just described is reversed, it is called a syphon. Such is the instrument described by *Bháskara Achárya* in his *S'iromani Mitákshara,* in the chapter on machines, where he says that if we place one end of such a curved tube, of copper or some other material, in a vessel full of water, the tube also being filled with water, and leave the other end open outside the vessel, all the water in the vessel will flow out through the tube.

m. In the same way if you use the hollow stalk of a lotus, the flower having been removed, a vessel of water may be emptied by it. Many ingenious machines are constructed on the principle of the syphon by conjurors and other artists.

n. The action of the syphon depends upon the weight of the atmosphere. The column of water in the external limb being longer than that in the limb immersed in the vessel, is heavier also. When it begins to fall, a vacuum would be formed in the upper part of the tube if the water in the shorter limb were not immediately forced to ascend, by the pressure of the atmosphere on the surface of the water in the vessel.

o. The air is the most common vehicle of sound, but not the only one. On the subject of Sound we proceed to make some remarks.

SECTION XI.

ACOUSTICS.

Aphorism LXVI.

Sound is heard when any sudden shock or impulse is given to the air, or to any body which is in contact, directly or indirectly, with the ear.

a. Thus sound, as that of a bell, can be heard under water.

b. Solid bodies also convey sound, as may be thus proved. Fasten a string, by the middle, round a bar of iron, and raise the bar from the ground by the two ends of the string, one end being held against each ear closed. If the bar then be struck with another piece of iron, the sound will be conveyed to the ear by means of the strings in a much more perfect manner than if it had no other vehicle than the air.

c. When a sonorous body, such as a bell, is struck, it is put into a state of rapid vibration. This vibration it communicates to the air, and the air communicates it to our ear, where it produces the sensation of sound.

d. A bell rung in a vessel from which the air has been removed, gives no sound.

Aphorism LXVII.

The tremulous motion given to the air by the vibration of a sonorous body resembles in some measure the motion communicated to smooth water when a stone is thrown into it.

a. A stone thrown into smooth water, first produces a small circular wave round the spot where the stone falls; the wave spreads, and gradually communicates its motion to the adjoining waters, producing similar waves to a considerable extent.

b. It requires some time for the vibrations of the air to extend to any distant spot. A washerman at some distance is seen to strike a mass of wet clothes upon a stone some time before the sound of the blow is heard.

Aphorism LXVIII.

The velocity of sound in air is computed to be at the rate of 1142 feet in a second.

a. The nearly uniform velocity of sound enables us to estimate the distance of the object from which it proceeds. If we do not hear the thunder till half a minute after we see the lightning, we infer that the cloud is at the distance of six miles and a half.

Aphorism LXIX.

An echo is produced when the aerial vibrations meet an obstacle.

a. When the waves raised in a tank, by throwing a stone into it, meet the bank, they are reflected. So an echo is produced when the aerial vibrations meet with an obstacle having a hard and regular surface, such as a wall or a rock. The vibrations may thus be reflected back to the ear, and produce the same sound a second time; but the sound will then appear to proceed from the object by which it is reflected ; just as the image reflected by a mirror seems to the eye to be behind the mirror.

b. If the vibrations fall perpendicularly on the obstacle, they are reflected back in the same line, if obliquely, the sound returns obliquely on the other side of the perpendicular, the angle of reflection, in this case as in others, being equal to the angle of incidence.

c. In solids, liquids, and even airs, we have found that *weight* presents itself, as a result of Attraction. But in the great cause of *Repulsion,* viz. Heat, though we find some properties which it has in common with the bodies already treated of, we do not find any evidence of weight.

SECTION XII.

OF THE IMPONDERABLES, HEAT, ELECTRICITY, &c.

Aphorism LXX.

Heat cannot be exhibited apart, nor be proved to have weight or inertia.

a. An iron ball, by being heated, does not gain in weight.

Aphorism LXXI.

Heat diffuses itself among neighbouring bodies until all have the same temperature.

a. A heated iron ball, thrown into a vessel of water, becomes cooled, while the water becomes heated.

M

Aphorism LXXII.

The inferior degrees of heat are denoted by the term *cold*.

a. A man who has descended half way from the summit of a pass in the Himálaya, exclaims—" How warm the air is!"— while the man who has just arrived at the same point from the plains, rejoins, " On the contrary—how very cold!" The air feels hot or cold to each of these two because its heat is greater or less than that of the air that he has left. Thus we learn that Sensation is not a perfect criterion of Heat.

b. As these two men give opposite evidence, from their feelings, so a man's two hands may give opposite evidence in regard to the heat of a basin of water. Put your right hand into as cold water as you can procure, and your left hand, at the same time, into as hot water as it can suffer, and, after a little time, place both hands in a basin of tepid water. To the right hand it will feel warm, and to the left hand cold. This shows still more clearly that Sensation is not a trustworthy thermometer.

Aphorism LXXIII.

Heat expands bodies, and hence any substance so circumstanced as to allow the expansion to be accurately measured, constitutes a Thermometer.

a. The common thermometer, for measuring the degrees of heat, is a glass bulb filled with mercury or other fluid, and having a narrow tube rising from it, into which the fluid, on being expanded by heat, ascends, and so marks the degree of heat.

Aphorism LXXIV.

Heat travels through different substances with different degrees of rapidity.

a. A burning match can be held till the flame comes near the fingers; but a metallic wire, if one end be put into the flame of a lamp, speedily becomes too hot to be held.

Aphorism LXXV.

Heat travels also through space.

a. It is then called Radiant Heat.

b. Heat proceeds from a heated body, such as an iron ball, in straight lines, and in all directions. The law of the equality of the angles of incidence and reflection applies to these rays. If the rays proceeding in this way from the Sun are received in a concave metallic mirror, they will be all reflected into one point, where the heat will be sufficient to set various substances on fire. Two such mirrors facing each other may be so arranged at opposite ends of a room that food placed in front of one of them shall be cooked by a fire placed in front of the other.

c. When a glass tube is rubbed with a piece of silk, or a large stick of sealing wax with a piece of flannel, then, besides heat, another effect is observed (—to which the name of *electrical* is given). The tube first attracts, and then repels light substances, such as feathers, and small pieces of straw or paper.

d. If the finger be brought near a large tube thus excited, a spark leaves the tube, and a slight crackling noise is heard. The same phenomenon occurs, in dry weather, if the hand is drawn rapidly over the back of a cat.

e. The sublime phenomenon of thunder and lightning is of the same nature as that just described. Lightning has been drawn from the clouds by means of a kite, and the same experi-

ments have been performed with it, as with the electricity deriv-
ed from the friction of glass and silk.

f. Electricity, like heat, always tends to an equilibrium.
Hence a cloud which by any cause has become highly electrified,
gives out its electricity with great noise and violence, when it
comes within a certain distance of another less highly electrified
cloud, or a tree, or a house.

g. When a body is highly electrified, its electricity can be
drawn away without noise or violence by bringing a sharp-point-
ed piece of metal near it. Advantage is taken of this fact, to
preserve houses from lightning in Calcutta and other places, by
attaching to the house a metallic rod, one end of which is buried
in the earth, and the other rises somewhat higher than the
house.

h. When substances are undergoing chemical changes, electrical phenomena appear. Electricity thus developed is called
Galvanism, from the name of its discoverer Galvani.

i. By means of Galvanism, information is conveyed in England in the course of a minute to places at the distance of hun-
dreds of miles. If the requisite arrangements were completed,
a message might be sent from Calcutta to Agra, Delhi, and
Bombay, within the hour.

j. By means of Galvanism also we can, although standing at
a great distance, instantaneously explode a barrel of gunpowder
at the bottom of a river, and thus remove the wreck of a vessel
which would otherwise obstruct the navigation of the river.
This has been lately done in the Ganges on several occasions.

k. Another effect of electricity is its giving to a piece of soft
iron the qualities of a magnet.

l. Light, also, is imponderable. The agency of Light in the phenomena of vision is the object of enquiry in the important department of science which we proceed next to treat of.

SECTION XIII.

OPTICS.

m. In this science, bodies are divided into *luminous, opaque,* and *transparent.* A luminous body is one that shines by its own light, as the sun, or a lamp. Polished metal, though it shines, is not luminous, for it would be dark if it did not receive light from a luminous body. It belongs to the class of opaque bodies which are neither luminous nor will allow the light to pass through them. Transparent bodies are those which allow the light to pass through them, such as glass and water. Transparent bodies are frequently called Mediums.

Aphorism LXXVI.

Light when emitted from a luminous body is projected in straight lines in every direction.

a. When the rays of light meet an opaque body, they are stopped short in their course. The interruption of the rays of light by the opaque body occasions darkness on the opposite side of it. If this darkness fall upon a wall or the like, it forms a shadow.

b. A shadow is not generally quite black, because it usually happens that light from another body reaches the space where the shadow is found, in which case the shadow is fainter. This happens if the opaque body be lighted by two lamps. If you extinguish one of them, the shadow will be both deeper and more distinct; yet it will not be perfectly dark, because it is still slightly illuminated by light reflected from the walls of the room, and other surrounding objects.

c. If the luminous body be larger than the opaque body, the shadow will gradually diminish in size till it terminate in a point. Thus the shadow of the moon, in an annular eclipse, terminates before it reaches the earth.

d. If the luminous body be smaller than the opaque body, the shadow will continually increase in size as it is more distant from the object that projects it. Thus the shadow of a man, thrown upon a wall by a lamp, may be twenty feet high or more.

Aphorism LXXVII.

Light moves with great velocity.

a. Light is about eight minutes and a half in its passage from the sun to the earth; therefore, when the rays reach us, the sun has quitted the spot he occupied on their departure; yet we see him in the direction of those rays, and consequently in a situation which he has abandoned about eight minutes and a half before.

b. The rate of the velocity of light was discovered in the following manner. We have already mentioned that the planet Jupiter is attended by four moons, which are very frequently eclipsed. When the earth is between Jupiter and the sun, then the earth is nearer to Jupiter, and when the sun is between the two, then the earth is further from Jupiter, by the distance of the diameter of the earth's orbit. Having ascertained exactly the time of the occurrence of the eclipses when Jupiter is near the earth, astronomers calculated, in accordance therewith, the time of many future eclipses; but it was found that, when the planet was at its greatest distance, the eclipses did not become visible till about 16 minutes after the calculated time. It was inferred that light takes this amount of time to traverse the earth's orbit, and from this the velocity of light was determined as follows.—As the sun's distance from the earth is about ninety five millions of miles, Jupiter, when at his greatest distance, must be a hundred and ninety millions of miles further off than when he is at his least distance from the earth; and it follows that the velocity of light is nearly two hundred thousand miles per second.

Aphorism LXXVIII.

When rays of light encounter an opaque body, after being stopped short, they are mostly reflected, as an elastic ball when flung against a wall.

a. Here as in other cases, the angle of reflection is equal to the angle of incidence. Admit a ray of the sun into a dark chamber by a very small hole, and let the ray fall perpendicularly on a mirror. Only one ray will be seen, for the incident and reflected rays are both in the same line, though in opposite directions. Hold the mirror so that the ray shall fall obliquely upon it; and the reflected ray will go off at the same angular distance on the other side of the perpendicular.

b. It is by reflected rays alone that we see opaque objects. Luminous bodies send rays immediately to our eyes; but the rays which they send to other bodies are invisible to us, and are seen only when reflected by those bodies to our eyes. The path of the ray which we spoke of as falling on the mirror, and reflected from it, was discernible only by means of the light reflected to the eye by small particles of dust floating in the air, on which the ray shone in its passage to and from the mirror.

Aphorism LXXIX.

When light passes from one into another transparent medium, it is refracted.

a. In consequence of this, when partly immersed in water, a staff appears shortened if it is placed vertically, and when it is placed obliquely the part immersed appears to be bent and broken off from the part above the water.

b. From the phenomena of refraction it would appear that when a ray of light passes from air into water, it is more strongly attracted by the water. If the ray fall perpendicularly on water, the attraction of the water acts in the same direction as the course of the ray. But if it fall obliquely, the water will attract it out of its course, and make it proceed in a direction more nearly vertical.

c. The converse is the case with a ray of light leaving a dense medium and entering a thinner one.

d. To show this,—place a silver coin at the bottom of a cup and set it so that the rim of the cup shall just intercept the rays of the sun or of a lamp and leave the coin in the shade. On filling the cup with water, the rays, being refracted downwards, will illuminate the coin.

e. Again, having emptied the cup replace the coin and move backwards till the rim of the cup shall hide the coin. Then let the cup be filled with water and the coin will become visible; for the rays reflected from the coin, and not intercepted by the rim of the vessel, which previously passed above the eye, being now refracted downwards, meet the eye.

f. For the same reason the bottom of a clear stream or lake appears more elevated than it really is. Ignorant boys, unable to swim, have lost their lives by going to bathe where the water did not appear to them to be deep.

g. The rays coming from the sun or from a star when near the horizon, are refracted downwards when they enter the atmosphere. The situation of the heavenly body in such circumstances therefore appears to be higher than it really is, for the eye perceives an object in the direction which the ray coming from it takes at the instant when it meets the eye.

h. The rays coming from a body in the zenith are not refracted.—The amount of refraction is greatest at the horizon, from which it gradually diminishes towards the zenith. Hence at sunrise the lower limb of the sun is more affected than the upper, and when the air is rendered more refractive by damp, the disk assumes a form approaching to that of an egg.

i. In astronomical observations, allowance must be made for refraction. Tables of its amount, at all points between the horizon, where it is greatest, and the zenith, where it ceases, have been constructed, and are printed in Astronomical treatises.

j. In passing obliquely through a pane of glass, a ray of light suffers two refractions. On coming from the thinner air into the denser glass, the refraction is towards a perpendicular to the inner surface of the glass. On passing from the denser glass into the thinner air, the refraction is equally away from a perpendicular to the outer surface. The two refractions therefore, be-

N

ing in contrary directions, produce nearly the same effect as if no refraction had taken place.

k. But this is the case only when the two surfaces of the refracting medium are parallel to each other : if they are not, the two refractions may be made in the same direction. Thus when parallel rays fall on a piece of glass having a double convex surface, which is called a *lens*, that only is perpendicular to the surface which falls in the direction of the axis, the axis being the straight line passing through the centre of the lens and perpendicular to both surfaces. All the other rays meeting the surface obliquely are reflected toward that perpendicular. When the rays quit the glass at the inner surface, they are again refracted towards the axis, by diverging from the perpendicular at the point where they leave the glass. Thus the whole eventually converge to a point behind the glass, called the *focus*, at about the distance of the radius of the sphere of which the surface of the lens forms a portion.

l. The sun's rays may be collected to a focus by a lens which, when used for this purpose, is called a burning-glass; and a piece of paper held in the focus may be set on fire.

m. A convex lens serves also as a magnifying glass.

n. The reason of this is as follows. An object at a distance from the eye appears small, and what is near appears large. A small object is seen with most distinctness, by a well constituted eye, when at the distance of about five inches. If if be brought nearer, it appears larger, but less distinct. This can easily be put to the test by holding the book at the distance first mentioned, and then nearer to the eye.

o. The reason of the indistinctness in the latter case is as follows. In the fore part of the eye there is a drop of pure liquid in the shape of a convex lens, the refractive power of which is such that rays diverging from a small object at the distance of about five inches are readily brought by it to a focus on the retina, where the concentration of rays produces distinct vision, but the refractive power is not sufficient to bring to a focus on the retina the rays diverging from a nearer object..

p. When people grow old, the fluid which forms the lens begins to diminish, and the convexity decreases, so that the letters of a book at the distance of five inches do not appear distinct to an old man. He sees more distinctly when he holds the book at arm's length.

q. This defect in the sight may be remedied in some measure by the use of convex spectacles, which cause the rays coming from any object to converge so far before reaching the eye that the lens of the eye can effect the rest.

r. By interposing a very refractive lens, a small object may be viewed distinctly when brought near the eye; and it then appears to be proportionately magnified. Such a glass is called a microscope.

s. A small convex lens may be obtained by piercing a small circular hole in a slip of metal, and introducing into it a drop of water, which will assume a spherical form on each side of the metal. Objects looked at through this will appear magnified.

t. If a lens be placed so as to fill up an aperture made in the window-shutter of a darkened room, and a sheet of paper be held at a proper distance behind it, then the objects outside, trees, houses, men walking, &c, will be represented on the paper. An arrangement analogous to this exists in the eye, as we shall show in the sequel.

u. The operation of a concave lens is the reverse of that of a convex lens. Objects seen through a concave lens are diminished.

v. In some young persons the lens of the eye is so highly convex that the rays come to a focus before they reach the retina.—From this focus the rays proceed divergently and consequently form a very confused image on the retina. Such persons see an object distinctly only when very near the eye.

w. Short sighted persons may enable themselves to see clearly at five inches distance by placing a concave lens before the eyes, in order to increase the divergence of the rays.

x. Short-sighted people may comfort themselves with the reflection that their eyes, by the flattening of the lens, may begin to improve, at a time of life when the sight of other people is beginning to fail.

y. There are other objects which though not really small, appear so to us, from their distance. In order to see these distinctly, we employ a Telescope.

z. The simplest form of Telescope consists of a convex lens fixed at the end of a tube, by which the rays from a distant object are made to converge; whilst the rays, before they come to a focus, are received on a concave lens which causes them to diverge so far as is necessary to give distinct vision. The object is thus seen magnified, and appears as if it were brought nearer.

Aphorism LXXX.

A beam of white light contains all the colours in nature.

a. Holding a clear glass bottle full of water higher than his head, or placing it on a support where a ray from the sun, coming into a dark room, may fall upon it, let the observer place himself between the sun and the bottle; then the various hues which appear in the rainbow, will all be exhibited by the bottle.* Let the bottle be raised or lowered until one of the coloured rays appears; then let the bottle be gradually raised, or lowered, and the other coloured rays will be seen in succession. Now how can such a variety of colours issue from clear water? The answer is this, that all these coloured rays are contained in a ray of white light. We must give some explanation of this.

a. It has been already mentioned that opaque bodies are seen by means of the rays which they reflect. All objects do not reflect every ray. A body which reflects all the rays appears white. That which reflects none appears black.

c. Along with the rays of light, the heat that attends them is either reflected or absorbed. Hence, in the sunshine, black clothes, which absorb the rays, are warm, and white clothes, which reflect them, are cool. Black cloth, for the same reason, takes fire quickly when exposed to a burning glass, but not so white cloth.

d. A body which reflects the red rays, appears red, that which reflects the yellow rays appears yellow, and so in every case.

e. The refrangibility of the coloured rays which make up white light is not equal. The blue rays are more refrangible than the yellow, and these than the red. This can be best shown by means of a prism of glass.

f. Let this be placed so that the ray of light entering a dark chamber by a small hole shall fall upon it, then the ray which

* This was the experiment of Antonio de Dominis, bishop of Spalatro, A. D. 1611.

first threw a white light on the opposite wall, will display the various colours of the rainbow. The refractions of the light, entering and quitting a prism, are both in the same direction, as in the case of a convex lens, and for the same reason; but, as its sides are flat, the rays are not brought to a focus.

g. If the coloured rays which have been separated by a prism are allowed to fall upon a convex lens, they will converge to a focus, where they will appear white as they did before refraction. Thus we can take a ray of white light to pieces, and put it together again.

h. The sun appears red through a fog, and also frequently at rising and setting, because the red rays, being less refrangible, reach our eyes, whilst the more refrangible rays do not.

i. The rainbow which exhibits a series of colours analogous to those of the prismatic spectrum, is formed by the refraction of the sun's rays in their passage through a shower of rain, every drop of which acts as a prism, in separating the coloured rays as they pass through it.

j. As all the drops which are placed at the same angle as regards the eye give the same colour, and the location of all the drops which are at similar angular distances from the eye must be in a circle, the form of the rainbow, of which we see only a portion, is necessarily circular.

k. "Inasmuch as the motion of bodies, the action of forces, and the propagation of influences of all sorts, take place in certain lines and over definite spaces, the properties of those lines and spaces are an important part of the laws to which those phenomena are themselves subject. Moreover, motions, forces or other influences, and times, are numerable quantities and the properties of number are applicable to them as to all other things."* Intending in the next section, therefore, to speak of Number and Magnitude, we here close the present Book.

* Mill's Logic—vol. 1. p. 394.

END OF THE SECOND BOOK.

भूमिका

सुनिपुणानां बुद्धिमतां विचारे परस्परविरोधः केवल दुःखहे-
तुः । वादिप्रतिवाद्यभिमतार्थस्वाभेदेऽपि यदि तयोर्भाषाभेदमा-
त्रेण भेदावभासः तर्हि सोऽपि तथैव । अन्योन्यमततत्त्वपरीच्छ-
णात्पूर्वं परस्परनिन्दादिकं निष्फलत्वादनुचितं । अपिच यच्च
केवलं विवदमानयोर्द्वयोरपि भ्रान्तिमूलकविवाददूरीकरणार्थः प्र-
यत्नो महाफलत्वाच्छस्तस्तच्च भूखण्डद्वयनिवासियावदुक्तीनां पर-
स्परं विवाददूरीकरणार्थकप्रयत्नः प्रशंसायोग्य इति किं वक्तव्यम् ।
एतादृग्प्रयत्नकारी पुरुषः सम्पूर्णफलस्याप्राप्तावपि न निन्द्यः ।
भारतवर्षीयार्यजनानां प्राचीनखमतग्रन्थपरिपालनं तत्प्रेमच तेषां
महास्तुतिकारणम् । एवं प्रतिदिनंवर्द्धमानखमतग्रन्थाभ्यासज-
नितसततज्ञानद्ध्या सन्तुष्यन्तो यूरोपीयलोका अपि न निन्द्याः ।
यदि कश्चिद् यूरोपीयजनो भारतवर्षीयार्येक्तां वास्तवमपि तदी-
यव्यवहारं तन्मततत्त्वं च यथार्थतो ऽविज्ञाय निन्देत्तदनुचितमेव ।
एवं यदि भारतीयजनो यूरोपीयमतमविज्ञाय निन्देत्तदपि तथैव ।
एवञ्च परस्परभ्रान्तिजनितमतविरोधप्रयुक्तदुःखस्य येयत्वा तद्-

दूरीकरणायावश्यं कश्चिदुपायो ऽनुष्ठातव्यः । मिथोविरुद्धेच मतद्वये सत्यता न सम्भवति । अनुचितमतस्वीकारे सति सत्फलासम्भवोऽनीप्सितदुष्टफलसम्भवश्च । अतो विचारिणोर्द्वयोरेकविषये मतभेदे सदसन्निर्णयाय वादः समुचितः । परन्तु याबत्सम्यक्प्रकारेण मतभेदो नावधृतस्तावद्वादोऽपि न समीचीनः । प्रथमतो मतयोः यथासम्भवं साम्यं निर्णीय तदुत्तरं भेदनिर्णयः कर्तव्यः येन मतैक्ये विवादो न भवेत् ॥

तदेवं पूर्वोक्तहेतुना भारतीयानां युरोपीयानांच मतयोस्साम्यवैलक्षण्यनिर्णयेच्छायां जातायां यस्मिन् ग्रन्थविशेषे भारतीयमतं सम्यक् साकल्येन स्पष्टीकर्तं ग्रन्थान्वेषणं प्राप्तं । तच्च तर्कशास्त्रीयग्रन्थानवलम्ब्य मतैक्यवैलक्षण्ये परीच्छणीये यतस्तेषु निखिलदर्शनस्थविषयाः संक्षेपेण बहवस्तच्छास्त्रीयविषया विस्तरेणोपलभ्यन्ते । भारतवर्षे बहूनां दर्शनानामुपलम्भेऽपि केवलं तर्कशास्त्रमात्रेण सह युरोपीयमतस्य साम्यवैलक्षण्ये निरूपणीये नत्वन्यैरपि तर्कशास्त्रज्ञानस्य सकलशास्त्रज्ञानमूलत्वात् । यथा प्रयागगयाकाशीयंत्रकस्य नगरत्रयस्य परस्परविप्रकर्षं तद्वस्थितिदिग्भागांश्च यो वस्तुतो जानाति स चेदुज्जायिन्यादेर्नगरान्तरस्य तत्त्रितयमध्ये एकेन केनचित्सह विप्रकर्षं दिश्च्च जानीयादपराभ्या-

मपि ताभ्यां तद्विषये निर्णयवान् स्यात् स्पष्टत्वात् । तथा यदि
सम्यक्परिचितसर्वदर्शनः पुरुषस्तर्कशास्त्रेण सह युरोपीयमतस्य
सम्मतिविमती जानाति तर्हि अन्यैरपि वेदान्तादिशास्त्रैः सह तस्य
साम्यवैलक्षण्यविषये गतसन्देहः स्यात् । तदेवं पूर्वोक्तविषयनिरू-
पणाय प्रारीप्सितस्य ग्रन्थस्याधिकारी स एव यो गीतमसूत्रप्रथ-
माध्यायार्थवेत्ताऽमत्सरस्य यतः प्रथमाध्याये गौतमेन सकलं न्या-
यमतं संक्षेपेण सूचितम् । मतयोर्वैलक्षण्ये परीक्षणीये प्रथमतो
विस्तरेणैक्यनिर्णयो यथासम्भवं कर्तव्यः यतस्तेषूपन्यस्तानां बहूना-
मुदाहरणानां वैलक्षण्यनिर्णये महानुपयोगो भविष्यति । अत्रात-
युरोपीयमतसम्बन्धिविषयाणां अतिस्पष्टरूपेण प्रकाशनार्थश्चैतद्
न्यनिर्माणमहाप्रयत्नः । तथाकृतेऽपि यदि श्रोतॄणां कस्मिंश्चिदृ-
ष्टाश्रुतपदार्थे सन्देहःस्यात् तर्हि तत्पदार्थनिरूपणनिपुणाद्युरो-
पीयजनान्निर्णयः कर्तव्यो न तु ग्रन्थ एव चेयतया ज्ञातव्यो ऽत्रा-
ततत्त्वे विषये बोधकापेक्षायाः सार्वत्रिकत्वात् । यः पुनर्विषयो
यन्त्रादिद्वारकदर्शनादेव निर्णेतुं शक्यस्तस्य निर्णयस्तेनैव
कर्तव्यो न तु तद्ग्रन्थवाक्यमात्रेण । यदि चाधीतात्मज्ञानशास्त्रः
प्रेक्षावानेतद्ग्रन्थद्वितीयप्रकरणे उपन्यस्तानां भौतिकपदार्थानां निरू-
पणं न रुच्या प्राधान्येन द्रष्यति तर्हि अनुमानीयदृष्टान्तसिद्ध्यर्थमेव

तत्प्रकरणं जानातु । यदि पुनरात्मतत्वज्ञो विविधयन्त्ररज्जुप-
भृतीन् तत्तद्व्यापारांश्चाच ग्रन्थे निरूपितानितितुच्छान्मन्येत तर्हि
यथैव पूर्वेषां ज्योतिःशास्त्राचार्याणां न्यायाचार्याणां वा ग्रन्थेषु
दृष्टान् तत्तद्यन्त्रादीन् दण्डचक्रचीवरकुम्भालकपालप्रभृतींश्च
जिज्ञासुजनबोधनार्थान् दृष्टान्तान्मन्यते तथैवैतानपि स्वीकरोतु
तत्तत्पदार्थस्वरूपपरिचये सति तदीयबहुविधापूर्वधर्मज्ञानरूपस्य
महाफलस्य सम्भवात् ॥

परमेश्वरकृपया चिरमर्थं बुध्वाच्चपादसूचाणां रचये
वृत्तिमपूर्वामिकुडीयाच्चपादतर्कैक्यां ॥

बहुलप्रयत्नरचिताबहुलार्थाल्पाचराप्यसन्दिग्धा बाल-
वुटैनस्य कृतिर्भूयादेषा मुदे विदुषां ॥

इह खल्वात्मादेः प्रमेयस्य तत्त्वज्ञानं परमपुरुषार्थो-
पकारीति सर्वे भारतवर्षीयास्तदन्यवर्षीयाश्चानुमन्यंते
शास्त्रकृतः। अतः परमपुरुषार्थार्थिभिरवश्यं तदात्मा-
दितत्त्वज्ञानप्राप्तौ यतितव्यं। तत्प्रासौच करणमपेक्ष्यत
इति तत्करणादीनि विचारयिष्यन् प्रमाणरूपकरणापु-
रसरं एतद्भन्यविषयानुद्दिशति प्रमाणेत्यादिना।

प्रमाणप्रमेयसंशयप्रयोजनदृष्टांतसिद्धांतवा-
यवतर्कनिर्णयवादजल्पवितंडाहेत्वाभासच्छ-
लजातिनिग्रहस्थानानां तत्त्वं विचार्यते त-
त्त्वज्ञानाय तस्य परमपुरुषार्थोपकारित्वात्
॥ १ ॥

___

अत्र चानुबंधाः विषयप्रयोजनसम्बन्धाधिकारिणः ।
विषयः प्रमाणादयः । प्रयोजनंचैतद्वन्याध्ययनप्राप्यं
तत्त्वज्ञानं पुरुषार्थप्रयोजकत्वात् । प्रतिपाद्यप्रतिपादक-
भावस्सम्बन्धः । उपदिष्टार्थशुश्रूषुरमत्सरोऽधिकारीति ॥
तत्त्वज्ञानं च क्रमेण परमपुरुषार्थोपकारीति क्रमप्र-
तिपादनाय सूत्रं ॥

___

दुःखजन्मकुप्रवृत्त्यनुचितरागादिमिथ्या-
ज्ञानानामुत्तरोत्तरापायमंतरा न परमः
पुरुषार्थः ॥ २ ॥

___

अस्यायमर्थः । अस्तिच तावत् मिथ्याज्ञानमनेकप्र-
कारकं । तस्माच्चानुचितरागादिकंभवति । आदिपदेना-
सूयादयो गृह्यंते । तस्माच्च जन्मकुप्रवृत्तिर्भवति । जन्म
जीवितं नतु पुनर्जीवितं लच्चयति । तस्मिन् जीवित-
काले अनुचितरागादिजन्या कुप्रवृत्तिः कुत्सिताचरणं
तस्माच्च दुःखमुत्पद्यते । अथ तल्लच्चनं मिथ्याज्ञानना-
श्रकं । मिथ्याज्ञानापायमंतरा न तज्जन्यरागादिरपैति ।
तदपायमंतरा न तज्जन्या कुप्रवृत्तिरपैति । तदपाय-
मंतरा न दुःखापायः । दुःखापायमंतरा न परमः पुरु-
षार्थः संभवतीति । इदंतु बोध्यम् । अनुचितरागादिकं
केवलमिथ्याज्ञानजन्यमस्तीति न । किन्तु मिथ्याज्ञानजन्यं
यद्यदनुचितरागादिकं तस्य तदापायमन्तरा अपायो
न भवति इति दिक् ॥

अथ यथोद्देशं लच्चणस्यापिच्चितत्वात् तल्लच्चानकरण-
भूतं प्रथमोद्दिष्टं प्रमाणं लच्चयति विभजतेच ॥

<hr>

॥ प्रत्यच्चानुमाने प्रमाणे ॥ ३ ॥

<hr>

४

अचच प्रमाकरणत्वं प्रमाणत्वं । प्रमाच यथार्थज्ञानं ।
लच्चितस्य विभाग: प्रत्यच्चानुमाने इति । अथ शब्देाप-
मानयोरपि प्रमाणत्वात् कथमुक्तं प्रमाणद्वयमेवेति चेदु-
च्यते । न तावच्छब्देापमाने अतिरिक्तप्रमाणे तयोर्लिंग-
विधया अनुमितिकरणत्वात् प्रमाणद्वयमेवेति सिद्धम् ॥
अथ विभक्ते यथाक्रमं लच्चयितुमारभते ॥

इंद्रियार्थसन्निकर्षोत्पन्नयथार्थ-
ज्ञानकरणं प्रत्यच्चं ॥ ४ ॥

इंद्रियस्यार्थेन सन्निकर्षात् उत्पद्यते यत् यथार्थं
ज्ञानं तत्करणं चच्चुरादि प्रत्यच्चप्रमाणमित्यर्थ: ॥
अनुमानं लच्चयति विभजतेच ॥

अथ तत्पूर्वकं त्रिविधमनुमानं पूर्ववच्छे-
षवत्सामान्यतो दृष्टंच ॥ ५ ॥

अथेति । तत्पूर्वकं प्रत्यच्चपूर्वकं त्रिप्रकारकमनुमान-
मित्यर्थ: । तस्य त्रैविध्यं दर्शयितुमाह पूर्ववदित्यादि ।

यच कार्येन कार्यमनुमीयते तत् पूर्ववत्। यथा मेघो
न्नत्या भविष्यति दृष्टिरिति। यच पुन: कार्येण कारण-
मनुमीयते तत् शेषवत्। यथा पूर्वौदकविपरीतमुद-
कं नद्या: पूर्णत्वं शीघ्रत्वंच स्रोतसो दृष्ट्वा ऽनुमीयते
भूता दृष्टिरिति। यचैकच पुष्पिताम्रदर्शनादन्यचापि
पुष्पिता आम्रा इति तत सामान्यतो दृष्टमिति॥

अथ शिष्यबोधनार्थमेवोपमानशब्दौ दर्शयिष्यन्ना-
दावुपमानं लक्षयति॥

प्रसिद्धसाधर्म्यात्साध्यसाधनमुपमानं॥ ६॥

प्रसिद्धस्य पूर्वंज्ञातस्य साधर्म्यात् साट्टश्यज्ञानात् सा-
ध्यस्य साधनं सिद्धिर्यंतस्तदुपमानमित्यर्थ:। यथा अयं
गोसदृश इति॥

शब्दं लक्षयति॥

आप्तोपदेश: शब्द:॥ ७॥

स्पष्टम्। विभजते॥

स द्विविधो दृष्टाट्टष्टार्थत्वात् ॥ ८ ॥

स शब्दो द्विप्रकारक इत्यर्थ: । यस्येह दृश्यते र्थस्स
दृष्टार्थ: । यथा गंगादिशब्द: । यस्य चामुच प्रतीयते
र्थ: सो ऽदृष्टार्थ: । यथा स्वर्गादिशब्द इति ॥

॥ समाप्तं प्रमाणप्रकरणम् ॥

अथानेन करणेन किं वेद्यमित्याकांचायां प्रमेयप्रक-
रणमारभमाण: प्रमेयं विभजते लच्चयतिच ॥

आत्मशरीरेन्द्रियार्थबुद्धिमन:-
कुप्रवृत्तिरागादिसांसारिकजीवनफ
लदु:खपरमपुरुषार्था: प्रमेयं ॥ ८ ॥

अत्रच आत्माच शरीरंच इंद्रियाणिच अर्थाश्च बुद्धिश्च
मनश्च कुप्रवृत्तिश्च रागादिच सांसारिकजीवनंच फलंच
दु:खंच परमपुरुषार्थश्चेति विग्रह: ॥
तत्र प्रथमोद्दिष्टमात्मानं लच्चयति ॥

इच्छाद्वेषप्रयत्नसुखदुःखज्ञानान्या-
त्मनो लिंगं ॥ १० ॥

लिंगं लच्चणमित्यर्थः ॥
क्रमप्राप्तं शरीरं लच्चयति ॥

चेष्टेंद्रियार्थोऽयंशरीरम् ॥ ११ ॥

चेष्टाच प्रयत्नजन्यो व्यापारविशेषः । चच्चुरादीनी-
न्द्रियाणि । अर्थसुखदुःखे । एतेषामाश्रयः शरीरमि-
त्यर्थः । अत्र चेष्टाश्रयत्वादि प्रत्येकं लच्चणं ॥
क्रमप्राप्तमिंद्रियं विभजते लच्चयतिच ॥

घ्राणरसनचच्चुस्त्वक्श्रोत्राणींद्रियाणि
भूतजन्यद्रव्यगुणग्राहकाणि ॥ १२ ॥

भूतानि वक्ष्यमाणानि । तज्जन्यगुणग्राहकाणीति स्व-
स्वविषयग्रहणलच्चणानींद्रियाणीत्यर्थः ॥

भूतान्येव कानीत्याकांच्चायामाह ॥

सुवर्णप्रभृतीनि भूतानि ॥ १३ ॥

सुवर्णममिश्रितं तदादीन्यमिश्रितानि भूतानीत्यर्थः ।
पृथिव्यादीनां भूतत्वमस्ति नवेत्यग्रे विवेचयिष्यते ॥
क्रमप्राप्तमर्थं विभजते लचयतिच ॥

गंधरसरूपस्पर्शशब्दा भूतजन्य
द्रव्यगुणास्तदर्थाः ॥ १४ ॥

तदर्थी इति तेषामिंद्रियाणामर्थाः । विषया इत्यर्थः ॥
बुद्धिं लचयति ॥

बुद्धिरुपलब्धिर्ज्ञानमित्यनर्थान्तरम् ॥ १५ ॥

अनर्थान्तरं समानार्थकमित्यर्थः ॥
मनो लचयति ॥

युगपाज्ज्ञानानुत्पत्तिर्मनसो लिंगं ॥ १६ ॥

युगपदेककाले । एकात्मनीति पूरणीयं । ज्ञानानाम-
नुत्पत्तिर्यतस्स एव धर्मो मनसो लिंगं लचणमित्यर्थः ॥

प्रवृत्तिं लच्चयति विभजतेच ॥

प्रवृत्तिर्बुद्धिशरीरारम्भः ॥ १७ ॥

उचितानुचितभेदेनापि प्रवृत्तेर्द्वैविध्यम् ॥

रागादि लच्चयति ॥

प्रवर्त्तनालच्चणं रागादि ॥ १८ ॥

प्रवर्त्तना प्रवृत्तिहेतुत्वं । तदेव लच्चणं यस्य ताट्टशं
रागादि । आदिपदार्थो वच्यमाण इति ॥
सांसारिकजीवनं लच्चयति ॥

ऐहिकदेहात्मसंयोगस्सांसारिकजीवनं ॥ १८ ॥

स्पष्टं ॥ फलं लच्चयति ॥

प्रवृत्तिजनितार्थः फलं ॥ २० ॥

प्रवृत्तेरुचितानुचितात्मिकायाः जनितोऽर्थसुखदुः-

ख

१०

खापभोगस्तत्फलमित्यर्थः ॥

दुःखं लच्चयति ॥

बाधनालच्चणं दुःखं ॥ २१ ॥

बाधना पीडा तद्देव लच्चणं यस्य ताट्टशमित्यर्थाः ॥

सकारणं परमपुरुषार्थं दर्शयति ॥

परमेश्वरप्रसादात्पर-
मः पुरुषार्थः ॥ २२ ॥

स्पष्टं ॥

॥ समाप्तं प्रमेयप्रकरणं ॥

अथ उद्देशलच्चणमाचेणात्मादिप्रमेयानामङ्गीका
रासं भवात्तदर्थमुपपत्तिर्निरूपणीया । साच संशय-
पूर्विकेति क्रमप्राप्तं संशयं लच्चयति ॥

सामान्यप्रत्यच्चाद्विशेषाप्रत्यच्चा-
द्विशेषस्मृतेश्च संशयः ॥ २३ ॥

सामान्यप्रत्यच्चात् सामान्यधर्मवज्ज्ञर्मिग्रहणात् । वि-
श्रेषाप्रत्यच्चात् विश्रेषाग्रहणात् । विश्रेषस्मृतेः विश्रेष
धर्मस्य स्थाणुत्वपुरुषत्वादेर्ग्रहणात्संशयो भवति ॥

अथ प्रयोजनाभावे संशयनिवृत्तये यत्नो न क्रियते ।
अतः क्रमप्राप्तं प्रयोजनं लच्चयति ॥

यमर्थमधिकृत्य प्रवर्त्तते तत्प्रयोजनं ॥ २४ ॥

अधिकृत्य उद्दिश्य । तथाच प्रवृत्तिहेत्विच्छाविषयः
प्रयोजनमित्यर्थः ॥

अथ दृष्टांताभावे सिद्धांताभावेवा उपपत्त्यसंभवात्
क्रमप्राप्तं दृष्टांतं लच्चयति ॥

लौकिकपरीच्चकाणां यस्मिन्नर्थे
बुद्धिसाम्यं स दृष्टांतः ॥ २५ ॥

अत्र लौकिकाः प्रतिपाद्याः । परीच्चकाः प्रतिपाद्काः ।
तेषां यस्मिन्नर्थे बुद्धिसाम्यं च्चानाविरोधा भवति स दृ-
ष्टांत इत्यर्थः । तथाच वादिप्रतिवादिनिश्चयविषयो

दृष्टांत इत्याशय: । यथा वह्निसाधने महानसं वह्लभा-
वसाधने महाह्रद इति ॥

॥ समाप्तं न्यायपूर्वाङ्गप्रकरणं ॥

क्रमप्राप्तं सिद्धांतं लच्चयति ॥

___

तंचाधिकरणाभ्युपगमसंस्थितिस्सिद्धांत: ॥ २६ ॥

___

तंचं शास्त्रं । तदेवाधिकरणं यस्य ताद्टशो यो अभ्युपगम-
स्तस्य संस्थितिरिदमित्थमिति व्यवस्था । तथाच शास्त्रि-
तार्थनिश्वयस्सिद्धांत इत्यर्थ: ॥

विभजते ॥

___

सर्वतंचप्रतितंचाधिकरणाभ्युपगमसं-
स्थितीनामर्थान्तरभावात् ॥ २७ ॥

___

सर्वतंचादिसंस्थितीनामर्थान्तरभावाङ्गेदादित्यर्थ: ।
तथाच चतुर्विधस्सिद्धांत इति फलितं ॥
सर्वतंचसिद्धांतं लच्चयति ॥

सर्वतंत्राविरुद्धस्तंत्रे ऽधिकृ-
तस्सर्वतंत्रसिद्धांत: ॥ २८ ॥

सर्वतंत्राविरुद्धस्सर्वशास्त्राविरुद्धस्तंत्रे प्रकृतशास्त्रे अ-
धिकृतस्सर्वतंत्रसिद्धांत: । यथा गंधादयो घ्राणादींद्रि-
यग्राह्या इति ॥

प्रतितंत्रसिद्धांतं लच्चयति ॥

समानतंत्रसिद्ध: परतंत्राऽसि-
द्ध:प्रतितंत्रसिद्धांत: ॥ २९ ॥

समान इति एकतंत्रसिद्ध: परतंत्राऽसिद्धस्तथाच वादि-
प्रतिवाद्येकतरमात्राभ्युपगत: प्रतितंत्रसिद्धांत इत्यर्थ: ।
यथेह्कुंडीयानां भूभ्रमणमिति ॥

अधिकरणसिद्धांतं लच्चयति ॥

यत्सिद्धावन्यप्रकरणसिद्धिस्सो
ऽधिकरणसिद्धांत: ॥ ३० ॥

यस्यार्थस्य सिद्धौ सत्यां अन्यस्यार्थान्तरस्य प्रक-

१४

रणात् सिद्धिर्भवति सोऽधिकरणासिद्धान्तः। यथा प्रप-
ञ्चस्य जन्यत्वे साध्यमाने परमेश्वरस्य सर्वज्ञत्वं॥

अभ्युपगमसिद्धांतं लचयति॥

अपरीचिताभ्युपगमात् तद्विशेषप
रीचणमभ्युपगमसिद्धांतः॥ ३१॥

अपरीचितस्य साचादसूचितस्य। अभ्युपगमादभ्युप-
गमन्नापकं यत्तद्विशेषपरीचणं भवति तस्माद्विशेषप-
रीचणात् चायते असूचितमभ्युपगतं सूचकृतेति सो
ऽयमभ्युपगमसिद्धांत इत्यर्थः। यथा गौतमेनाभ्युपगतं
मनस इंद्रियत्वमिति॥

॥ समाप्तं न्यायाश्रयसिद्धांतप्रकरणं॥
क्रमप्राप्तानवयवान् लचयितुं विभजते॥

प्रतिज्ञाहेतूदाहरणोपनय-
निगमनान्यवयवाः॥ ३२॥

अत्र प्रतिज्ञादयः पंचन्यायावयवा इत्यर्थः। पंचावय-

वक्तथनं पराथीनुमानाभिप्रायकं ॥

प्रतिज्ञां लच्चयति ॥

साध्यनिर्देशः प्रतिज्ञा ॥ ३३ ॥

साधनीयस्यार्थस्य यो निर्देशस्सप्रतिज्ञावयव इत्यर्थः ॥
यथा भूगोलाकृतिरिति ॥
क्रमप्राप्तं हेतुं लच्चयति विभजतेच सूचाभ्याम् ॥

उदाहरणसाधर्म्यात्साध्यसाधनं
हेतुः तथा वैधर्म्यात् ॥ ३४ ॥

अत्र साध्यसाधनं हेतुरिति सामान्यलच्चणं । तस्य द्वै-
विध्यमाह उदाहरणसाधर्म्यात् तथा वैधर्म्यादिति । सा-
धर्म्यमन्वयः । ताट्टश्यव्याप्तिरित्यर्थः । तथा चान्वयव्या-
प्तिकहेतुबोधको हेत्ववयव इत्यर्थः । यथा उक्तसाध्ये
नियतगोलच्छायावच्चात् । यत् नियतगोलच्छायकं तत्
गोलं दृष्टं यथा कंदुकमिति । वैधर्म्यं व्यतिरेकः । तथाच
व्यतिरेकव्याप्तिहेतुबोधको हेत्ववयव इत्यर्थः । यथा उक्त-

साध्ये नियतगोलच्छायावत्त्वात् । यन्नोलं न भवति न
तत् नियतगोलच्छायकं यथा स्तंभ इति ॥

क्रमप्राप्तमुदाहरणं लच्चयति ॥

साध्यसाधर्म्यात्तद्धर्मभावी दृष्टांत उदाहरणं ॥ ३५ ॥

साध्यसाधर्म्यात् साध्यसहचरितधर्म्मात् । तद्धर्मभावी-
ति तं साध्यरूपं धर्मं भावयतीति । तथाच साधनवत्ता-
प्रयुक्तसाध्यवत्तानुभावकोऽवयव इत्यर्थः । यथा यद्वत्
नियतगोलच्छायकं तद्गोलाकृतिकं यथा कंदुकमिति ॥

व्यतिरेक्युदाहरणं लच्चयति ॥

तद्विपर्ययाद्वा विपरीतं व्यतिरेक्युदाहरणं ॥ ३६ ॥

तद्विपर्ययादिति साध्यसाधनव्यतिरेकव्याप्तिप्रद-
र्शनात् । तथाच साध्यसाधनयोर्व्यतिरेकव्याप्त्युपदर्शकी-
दाहरणं व्यतिरेक्युदाहरणमित्यर्थः । यन्नोलं न भवति
न तत् नियतगोलच्छायकं यथा स्तंभ इति वाकारः प्र-
योगसूचकः ॥

क्रमप्राप्तमुपनयं लक्षयति ॥

उदाहरणापेच्चस्तथेत्युपसंहारो न त-
थेति वा साध्यस्योपनयः ॥ ३७ ॥

साध्यस्य पच्चस्य उदाहरणापेच्च: उदाहरणानुसारी
य उपसंहारः उपन्यासः स उपनयावयव इत्यर्थः । स
द्विविधोऽन्वयिव्यतिरेकिभेदात् । तथेति साध्यस्योपसं-
हारो ऽन्वयुपनयः । यथा तथाचेयं भूः । न तथेति
साध्यस्योपसंहरो व्यतिरेकुपनय इति ॥
निगमनं लक्षयति ॥

हेत्वपदेशात्प्रतिज्ञायाः पुनर्वचनं निगमनं ॥ ३८ ॥

हेतोर्थात्तिविशिष्टपच्चधर्मस्य अपदेश: कथनं त-
त्पूर्वकं प्रतिज्ञायाः पुनर्वचनं निगमनावयव इत्यर्थः ।
यथा तस्मात्तथालाकारा भूरिति ॥

॥ समाप्तं न्यायस्वरूपप्रकरणं ॥

अथ प्रतिवादिनो निर्णयप्रतिबन्धकहेतौ व्यभिचा-
रशंकानिरासार्थं तर्कस्यावश्यकत्वात् निर्णयात्पूर्वं क्रम-
प्राप्तं तर्कं लक्षयति ॥

अविज्ञाततत्त्वेऽर्थे कारणोपपत्तित-
स्तत्त्वज्ञानार्थमूहस्तर्कः ॥ ३८ ॥

अत्र तर्क इति लक्ष्यनिर्देशः । ऊह इति लक्षणं । ऊह-
स्व तर्कयामीत्यनुभवसिद्धो मानसो धर्मः । तर्कः किं
स्वत एव निर्णायकः परंपरयावेत्यत आह कारणेति ।
कारणस्य व्याप्तिज्ञानादेरुपपादनद्वारेत्यर्थः । तथाच
नियतगोलच्छाया यदि गोलव्यभिचारिणी स्यात् तर्हि
गोलजन्या न स्यादित्यनेन व्यभिचारशंकानिरासे निरा-
बाधं व्याप्तिज्ञाने न भुवो गोलत्वानुमितिरिति ॥
क्रमप्राप्तं निर्णयं लक्षयति ॥

विमृश्य पक्षप्रतिपक्षाभ्यामर्थावधा-

रणं निर्णय: ॥ ४० ॥

विमृश्य संदिग्धे पच्चप्रतिपच्चाभ्यां स्वपच्चस्थापनप-
रपच्चदूषणाभ्यां अर्थावधारणं निर्णय इत्यर्थः ॥

॥ समाप्तं न्यायोत्तराङ्गप्रकरणं ॥

॥ इति प्रथमस्याध्यायस्याद्यमाह्निकं ॥

अथ तर्कानंतरं साधुस्तत्त्वजिज्ञासुर्वादमंतरा कदा-
चिन्नर्थयं न प्राप्नोतीति क्रमप्राप्तं वादं लच्चयति ॥

प्रमाणतर्कसाधनोपालंभ: सिद्धांताविरुद्ध: पंचावय-
वोपपन्नपच्चप्रतिपच्चपरिग्रहो वाद: ॥ ४१ ॥

अचच वाद इति लच्चनिर्देशः । एकाधिकरणस्थौ वि-
रुद्धौ धर्मौ पच्चप्रतिपच्चौ । तयो: परिग्रहस्तत्साधकोक्ति
प्रत्युक्तिरूपवचनसंदर्भ इत्यर्थः । प्रमाणतर्कसाधनोपालं-
भ: प्रमाणैस्तर्केणच साधनं स्थापनं तथा उपालंभ: निषे-
धश्च यस्मिन् ताटृश इत्यर्थः । सिद्धांताविरुद्धस्सिद्धांता-

नुकूलः । पंचावयवोपपन्नः पंचावयवयुक्त इत्यर्थः ॥
जिगीषयाहि धूर्त्तास्त्वजिज्ञासाच्छलेन जल्पं कुर्व-
न्तीति क्रमप्राप्तं जल्पं लच्चयति ॥

यथोक्तोपपन्नः छलजातिनिग्रहस्था-
नसाधनोपलंभो जल्पः ॥४२॥

यथोक्तोपपन्न इति अनेनप्रमाणतर्कसाधनोपालंभः
पच्चप्रतिपच्चपरिग्रह इति लभ्यते । छलजातिनिग्रह-
स्थानैस्साधनोपालंभौ स्थापनानिषेधौ यस्मिन् क्रियेत
इति । तथाचोभयपच्चस्थापनावती विजिगीषुकथा
जल्प इत्याशय: ॥

धूर्ता अपि जल्पासमर्था वितण्डां कुर्वंतीति क्रमप्रा-
प्तां वितंडां लच्चयति ॥

स प्रतिपच्चस्थापनाहीनो वितण्डा ॥ ४३ ॥

स जल्प: । प्रतिपच्चो द्वितीय: पच्च: । तथाच प्रति-
पच्चस्थापनाहीना विजिगीषुकथा वितण्डेत्यर्थ: ॥

॥ समाप्तं कथाप्रकरणं ॥

अथ जल्पकादयस्खमतनिर्णये हेतुमप्राप्य हेत्वाभा-
सान् प्रयुंक्ते । तथा भ्रमात्साधवोऽपीति क्रमप्राप्तान्
हेत्वाभासान् लच्चयति विभजतेच ॥

सव्यभिचारविरुद्धप्रकरणसमसाध्यसमा-
तीतकाला हेत्वाभासाः ॥ ४४ ॥

सव्यभिचारश्च विरुद्धश्च प्रकरणसमश्च साध्यसमश्च
अतीतकालश्चेति विग्रहः । हेतुवदाभासन्त इति हेत्वा-
भासा दुष्टहेतव इत्यर्थः ॥

सव्यभिचारं लच्चयति ॥

अनैकान्तिकस्सव्यभिचारः ॥ ४५ ॥

एकस्य साध्यस्य तदभावस्यवा योऽन्तस्सहचारस्त-
थाचैकमाचव्याप्तिग्राहकस्सहचारस्वेकांतिकस्तदन्योऽनै-
कांतिकः । यथा पर्वतो धूमवान् वह्नेरिति । सच

साधारणादिभेदात्त्रिविध इति ॥

क्रमप्राप्तं विरुद्धं लच्चयति ॥

———————

सिद्धान्तमभ्युपेत्य तद्विरोधी विरुद्ध: ॥ ४६ ॥

———————

सिद्धांतं साध्यमभ्युपेत्य उद्दिश्य प्रयुक्त: तद्विरोधी
साध्याभावव्याप्त इति फलितार्थ: ।   यथा वह्निमान
ह्रदत्वादिति ॥

क्रमप्राप्तं प्रकरणसमं लच्चयति ॥

———————

यस्मात्प्रकरणचिन्ता स निर्णयार्थ-
मपदिष्ट: प्रकरणसम: ॥ ४७ ॥

———————

स हेतुस्खसाध्यस्य परसाध्याभावस्य वा निर्णयार्थम-
पदिष्ट:प्रयुक्त: प्रकरणसम उच्यते । सक इत्याकांच्चाया-
माह यस्मादिति । प्रकरणं साध्यतदभाववंतौ । तथाच
निर्णयार्थं प्रयुक्तो हेतुर्यच निर्णयं जनयितुमशक्त: किं-

त धर्मिणस्साध्यवलं तदभाववलंवेति चिन्तां प्रवर्त्तयति
स प्रकरणसम इत्युच्यते । यथा शब्दो नित्यश्शब्दत्वात् ।
शब्दो ऽनित्य: कृतकत्वादिति ॥

क्रमप्राप्तं साध्यसमं लच्चयति ॥

साध्याविशिष्टश्च साध्यत्वात्साध्यसम: ॥ ४८ ॥

साध्येनाविशिष्ट: कुत इत्यत आह साध्यत्वादिति ।
साधनीयत्वादित्यर्थः । यथाहि साध्यं साधनीयं तथा हेतु-
रपिचेत् साध्यसम इत्युच्यते । अत एवचासिद्ध इति व्यव-
ह्रीयते । अयंचाश्रयासिद्ध्यादिभेदेन त्रिविध इति ॥

क्रमप्राप्तमतीतकालं लच्चयति ॥

कालात्ययापदिष्ट: कालातीत: ॥ ४९ ॥

कालस्य साधनकालस्यात्यये अभावे अपदिष्ट: अ-
युक्तो हेतु: । एतेन साध्याभावप्रमालच्चणार्थं इति सू-
चितं । साध्याभावनिर्णये साधनासंभवात् अयमेव बा-

धितसाध्यक इति गीयते । यथा वह्निरनुष्णः कृतक-
त्वादिति ॥

॥ समाप्तं हेत्वाभासप्रकरणं ॥

हेत्वाभासान् साधुरपि भ्रमादेव प्रयुंक्ते इत्युक्तं । अथ
धूर्त्तमाचप्रयुक्तं क्रमप्राप्तं छलं लचयति ॥

वचनविघातो ऽर्थविकल्पोपपत्त्याच्छलं ॥ ५० ॥

अर्थस्य वाच्यभिमतस्य यो विकल्पो विरुद्धः कल्पो अ-
र्थान्तरकल्पनेति यावत् तदुपपत्त्या युक्तिविशेषेण यो
वचनस्य वाच्यक्रस्य विघातो दूषणं छलमित्यर्थः । वक्तृ-
तात्पर्यीविषयार्थकल्पनेन दूषणाभिधानमिति फलि-
तार्थः ॥

छलं विभजते ॥

तद्त्रिविधं वाक्छलं सामान्यच्छलमुप-
चारच्छलंच ॥ ५१ ॥

वाक्छलादिभेदात् त्रिविधं छलमित्यर्थः ॥
वाक्छलं लच्चयति ॥

___

अविशेषाभिहितेऽर्थे वक्तुरभिप्रायाद-
र्थान्तरकल्पना वाक्छलं ॥ ५२ ॥

___

यत्र शब्दस्यार्थद्वयं तत्राविशेषेण एकतरानिश्चये
न अभिहितेऽर्थे कथिते सति वक्तुरभिप्रायादर्थान्तरस्य
वाद्यनभिमतस्य कल्पनया दूषणाभिधानं वाक्छल-
मित्यर्थः । यथा गौर्विषाणीत्युक्ते कुतो गजस्य भृंग-
मिति ॥

सामान्यछलं लच्चयति ॥

___

सम्भवतोऽर्थस्यातिसामान्ययोगादसम्भू-
तार्थकल्पना सामान्यछलं ॥ ५३ ॥

___

जातिसम्बन्धात् संभवदर्थाभिप्रायेणोक्तस्य अति
सामान्ययोगादसंभवदर्थकत्वकल्पनया दूषणाभिधानं
सामान्यछलं । यथा ब्राह्मणोऽयं विद्याचरणसंपन्न

इत्युक्ते कथमस्य विद्याचरणसंपन्नत्वं बाल्ये व्यभिचारा-
दिति ॥

उपचारछलं लच्चयति ॥

धर्मविकल्पनिर्देशे ऽर्थसद्भावप्रतिषेध
उपचारछलं ॥ ५४ ॥

धर्मविकल्पोऽन्यच दृष्टस्यान्यच प्रयोग: तस्य निर्देशे
अर्थसद्भावप्रतिषेध उपचारछलं । यथा मञ्चा: क्रोशं-
तीत्युक्ते मंचस्थ पुरुषा: क्रोशंति नतु मंचा इति ॥

प्रसंगात् छलं पूर्वपच्चयति ॥

वाक्छलमेवोपचारछलं तदविशेषात् ॥ ५५ ॥

शब्दस्यार्थान्तरकल्पनया तयोरविशेषात् वाक्छल-
मेवोपचारछलं । तथाच द्विविधमेव छलमित्यर्थ: ॥

समाधत्ते ॥

न तदर्थान्तरभावात् ॥ ५६ ॥

वाक्छलमेवोपचारछलं न भवति तयोरर्थांतर भा—
वाङ्गिन्नत्वादित्यर्थः ॥

विपच्चे बाधकमाह ॥

अविशेषेणा किंचित्साधर्म्यादेकच्छलप्रसंगः ॥ ५७ ॥

यल्किंचिड्धर्मादविशेषे यल्किंचित्साधर्म्यात् छलत्वा—
दिरूपाच्छलैकयं स्यात् नतु त्वदभिमतं द्विलमपीति ॥

॥ समाप्तं छलप्रकरणम् ॥

क्रमप्राप्तां जातिं लच्चयति ॥

साधर्म्यवैधर्म्याभ्यां प्रत्यवस्थानं जातिः ॥ ५८ ॥

साधर्म्यवैधर्म्याभ्यां प्रत्यवस्थानं दूषणाभिधानं जातिरि—
त्यर्थः । तथाच छलभिन्नदूषणासमर्थमुत्तरं स्वव्याघात—
कमुत्तरं वा जातिरिति सूचितं ॥

क्रमप्राप्तां निग्रहस्थानं लच्चयति ॥

विप्रतिपत्तिरप्रतिपत्तिश्च निग्रहस्थानं ॥ ५९ ॥

विपरीता प्रतिपत्तिर्विप्रतिपत्तिः परस्थापितात्प्रति-
षेधः प्रतिषेधानुद्धरणं वा अप्रतिपत्तिः उभयंच निग्र-
हस्थानं पराजयप्राप्तिः ॥

जातिनिग्रहस्थानयोर्विभागो नास्तीति भ्रमवारणा-
याह ॥

तद्विकल्पाज्जातिनिग्रहस्थानबहुत्वं ॥ ६० ॥

तयोर्विप्रतिपत्त्यप्रतिपत्त्योर्विकल्पाद्भेदाज्जातिनिग्र-
हस्थानानां बहुत्वं । दूर्त्यंच तयोर्बहुत्वेऽपि अन्य वि-
षयकमिष्यजिज्ञासायाः प्रतिबंधान्नेदानीं तद्विभागः
क्रियत इति भावः ॥

इति श्रीमज्जेम्स् बालण्टैनविरचितायां न्यायसू-
चकौमुद्यां प्रथमोऽध्यायस्समाप्तः ॥

अथ प्रथमाध्याये प्रत्यक्षकरणानामिन्द्रियाणां तद-
र्थानाञ्च उद्देशः कृतः । इदानीं विशेषेण तल्लक्षणानि
दर्शयितव्यानि । तत्र प्रथमोद्दिष्टं घ्राणेन्द्रियं लक्ष-
यति ॥

॥ गंधग्राहकमिन्द्रियं घ्राणं नासावर्त्ति ॥ १ ॥

गंधग्राहकमिन्द्रियं घ्राणेन्द्रियमित्यर्थः । ननु कथं
तस्य गंधग्राहकत्वं इति चेदुच्यते । गंधवत् पदार्थांशाः
श्वासद्वारा नासां प्रविश्य यदा घ्राणेंद्रियेण सह संयुज्ल्ते
तदैव प्रतिबन्धकाभावे गन्धज्ञानमुत्पद्यते । इति तस्य-
गन्धग्राहकत्वं ॥ याट्टशशक्तिसत्त्वेनैव पदार्थांशानां घ्रा-
णेन्द्रियेण सह संयोगे सति गंधज्ञानमुत्पद्यते ताट्टश-
शक्तिमत्पदार्थो गंधवान् बोध्यः । शक्तिनिरूपणंचाग्रे-
स्फुटीभविष्यति ॥ क्रमप्राप्तं रसनेन्द्रियं लक्षयति ॥

॥ रसग्राहकमिन्द्रियं रसनं जिह्वावर्त्ति ॥ २ ॥

रसग्राहकमिंद्रियं रसनेंद्रियमित्यर्थः । कथं तस्य र-
सग्राहकत्वं इति चेदुच्यते । मुखप्रविष्टा रसवत्पदार्था
मुखजलसंयोगेन द्रवीभावं प्राप्ता यदा रसनेंद्रियसंयु-
क्ता भवन्ति तदैव प्रतिबन्धकाभावे रसज्ञानमुत्पद्यते इति
तस्य रसग्राहकत्वं । याट्टग्शक्तिसत्त्वेनैव रसनेंद्रियसंयु-
क्तपदार्थगतरसप्रत्यक्षं भवति स एव पदार्थो रसवानि-
त्यभिधीयते ॥ क्रमप्राप्तं चक्षुरिन्द्रियं दर्शयति ॥

॥ रूपग्राहकमिन्द्रियं चक्षुर्नेत्रवर्त्ति ॥ ३ ॥

रूपग्राहकमिंद्रियं चक्षुरिंद्रियमित्यर्थः । कथं तस्य
रूपग्राहकत्वं इति चेदुच्यते । स्वतःप्रकाशात् परतःप्र-
काशाद्वा पदार्थादागतं यत्तेजो यदा नेत्रं प्रविशति
तदा तत्तेजःसम्बद्धस्यरूपस्य चाक्षुषज्ञानमुत्पद्यते । तेजः
स्वरूपं तेजोरूपयोः सम्बन्धः नेत्रघटकावयवव्यवस्था-

चाग्रे विचारणीया ॥

क्रमप्राप्तं त्वगिन्द्रियं दर्शयति ॥

स्पर्शग्राहकमिन्द्रियं त्वक् सर्वशरीरवर्त्ति अंगु-

ल्यग्रेषु तस्योत्कर्षः संख्यापरिमाणादीनां तत

एव विशेषावगमात् ॥ ४ ॥

स्पर्शग्राहकमिन्द्रियं त्वगिंद्रियमित्यर्थः ॥ क्रमप्राप्तं

स्ववर्णेंद्रियं लच्चयति ॥

कर्णवर्त्तींन्द्रियं श्रोचं ॥ ५ ॥

अथैतेनेंद्रियेण किं चायते । उच्यते । अवणानुकूल

कर्णवर्त्तिंत्वच: स्पंदनात् शब्दो चायते । शब्दस्य स्व-

रूपं कर्णघटकावयवव्यवस्थाचाग्रे विचारणीया ॥

द्वतीन्द्रियनिरूपणं समाप्तं ॥ अथेदानीमिंद्रियाणा-

मर्था दर्शनीयाः । तच्च प्रथमोद्दिष्टं गंधं दर्शयति ॥

याट्टश्यभक्त्या द्रव्यांशानां आर्घेद्रियेण सह संयो-
गे सति गंधज्ञानमुत्पद्यते सेव भक्तिर्गन्ध इ-
त्यभिधीयते ॥ ६ ॥

यद्गन्धवत् तत् सर्वं पृथिवीति नियमे माना-
भावात् फलाभावाच्चैतन्नियमं त्यक्त्वायस्मिन् यस्मिन
द्रवद्रव्ये अद्रवद्रव्ये वा गंध उपलभ्यते तत् गंधाश्रय
इति व्यवह्रीयते । एतद्विषये किंचिदग्रे विचारयिष्यते ॥

क्रमप्राप्तं रसं निरूपयति ॥

याट्टश्यभक्त्या द्रव्यांशानां रसनेन्द्रियसंयोगे सति रस-
ज्ञानमुत्पद्यते सा भक्तीरस इत्युच्यते ॥ ७ ॥

अस्मन्मते स्वच्छजलस्यापि न रसवत्त्वं किन्तु मुख-
सम्बन्धिजलेन रसवद्द्रव्ये द्रवीभूते सति इन्द्रियसन्नि-
कर्षद्वारा रसप्रत्यक्षं भवतीति रसप्रत्यक्षे जलस्योप-
योगित्वमिति ॥

क्रमप्राप्तं रूपं निरूपयति ॥

चक्षुपा तेजःसंयोगे सति याट्टश्यशक्त्या
रूपप्रत्यक्षं भवति सा शक्तीरूपं दृ-
त्यभिधीयते ॥ ८ ॥

तेजस्तत्तद्द्रव्यसंयोगजन्या अवस्थाविशेषा अग्रे वि-
चारणीयाः ॥

अथ चक्षुरिन्द्रियं खत एव पदार्थदूरत्वव्यापकं न
वेत्याकांक्षायां सूत्रम् ॥

यत्पदार्थस्य न स्पष्टरूपादिज्ञानं तस्य
दूरत्वमदृश्यमस्पष्टरूपादिज्ञानाच्चस्य
तदनुमेयम् ॥ ९ ॥

तस्य पदार्थस्य दूरत्वमदृश्यं चाक्षुषप्रत्यक्षाविषय
इत्यर्थः । नन्वेवं सति तत् कथं ज्ञायत इत्याकांक्षाया-
माह अनुमेयमिति । अनुमित्यात्मकज्ञानविषय इत्य-
र्थः । तत्र हेतुमाह अस्पष्टरूपादिज्ञानादिति । त-
थाहि । यत्समीपस्थं तत् प्रायः स्पष्टं दृष्यते यच्च न

ङ

तथा तदस्पष्टं दृश्यते । यथा । अतिसच्चगवाच्चद्वारा
मद्वारच्चस्याप्यस्पष्टत्वेन ज्ञानम्भवति ताद्दृशास्पष्टत्वज्ञा-
नात् तस्य दूरत्वमनुमेयमेवेति । अत एव यद्यपि सू-
र्यस्य दूरत्वं पञ्चचत्वारिंशन्नियुतक्रोशपरिमितमस्ति त-
थापि अतिस्पष्टदर्शनात् तस्य तन्नानुमीयते किन्तु कुम्भ-
कारचक्रात्तस्याल्पतरपरिमाणं चक्षुषा गृह्यत इति ॥

अस्पष्टत्वादित्यचादिपदात् पूर्वज्ञातस्य महत्परि-
माणविशिष्टपदार्थस्य अल्पत्वेन ज्ञानं तज्ज्ञानकारणा-
ज्ञानञ्च । यथा । मिहिकायां सत्यां सायङ्कालिक-
तमसि वा समीपवर्त्ती काकादि: दूरवर्त्ती हस्तीव
गृह्यते अल्पत्वज्ञानकारणमिहिरान्धकारादेरननुस-
न्धानादिति ॥

क्रमप्राप्तं स्पर्शं निरूपयति ॥

द्रव्ये त्वगिन्द्रियसंयोगे सति याद्दृश-
श्रक्त्या स्पार्शनप्रत्यक्षमुत्पद्यते सा
श्रक्ति: स्पर्श: ॥ १० ॥

क्रमप्राप्तं शब्दं निरूपयति ॥

याट्टशशक्त्या द्रव्यविशेषस्य स्पन्दनेन
श्रावणप्रत्यच्चानुकूलत्वक्स्पन्दनमुत्प-
द्यते सा शक्ति: शब्द इत्युच्यते ॥ ११ ॥

यथाच शब्दोत्पत्तिस्थानविशेषद्रव्यस्पन्दनस्य वाय्वा-
दिद्वारा कर्णवर्त्तित्वक्स्पन्दनोत्पादकस्य शब्दोत्पादकत्वं
तथाग्रे वक्ष्यते ॥

अथ श्रोचेन्द्रियं खत एव पदार्थदूरत्वज्ञापकं नवे-
त्याकांच्चायां सूचम् ॥

शब्दोत्पादकपदार्थस्य दूरत्वं न श्रूयते
किन्तु तत्तच्छब्दस्य महत्वलघुत्वानु-
सारेण तत्तच्छब्दकारणदूरत्वसमीप-
त्वज्ञानानुसारेण चानुमीयते ॥ १२ ॥

शब्दस्येति । दूरत्वं न श्रूयते न श्रवणेन्द्रियजन्य-
प्रत्यच्चविषयीक्रियत इत्यर्थ: । अथ कथं तर्हि त-

ज्ञानमित्यत आह किन्निति । अनुमीयत इति ।
अनुमानप्रमाणविषयीक्रियत इत्यर्थः । तथाहि ।
कश्चिन्निरिनदीधारापतनदेशसमीपस्थस्तत्पतनशब्द प-
रिमाणं निश्चित्य ततः क्रमशोऽपसृत्य अपसरणदेशदू-
रत्वानुसारेण पूर्वशब्दमल्पपरिमाणकं शृणोति । त-
दनु तत्सौच्म्येन प्रायशो यथार्थमेव धारापतनदेश-
दूरत्वमनुमातुं शक्नोति । एवमेव यदा शब्दकार-
णमतिदूरष्टत्तीति ज्ञायते तदा ज्ञातपरिमाणापेच्चया
महत्परिमाणं शब्दस्य स्यादित्यप्यनुमातुं शक्नोति ।
अत एव यः कश्चित्पञ्चक्रोशपरिमितदेशादागतः शब्दः
सन्निहितनलिकाशब्दाद्प्यल्पतरोऽपि शतघ्नीशब्द ए-
वायमिति तत्कारणदूरत्वचैरनुमीयते । यदिच ध्व-
निकारणस्य दूरत्वं सामीप्यंवा न जानीयात्तर्हि दूरष्ट-
त्तिमेघस्य अथवा समीपवत्तिरथस्यायं ध्वनिरिति शं-
का भवेत् तस्मात् दूरत्वमनुमेयमेवेति सिद्धम् ॥

अथैवं निरूपिता ये रूपाद्यो गुणास्तद्वन्ति द्रव्याणि
प्रत्येकं कतिविधानि इत्याकांचायां सूचम् ॥

वह्निरिन्द्रियग्राह्यगुणवद्द्रव्याणि कार्यकारणगत-

महत्त्वाणुत्वधर्मभेदात् प्रत्येकं द्विविधानि ॥ १३ ॥

बहिरिन्द्रियग्राह्या ये रूपादयो गुणास्तद्वन्ति द्रव्याणि
द्विविधानीत्यर्थः । कथं तेषां द्विविधं द्रव्यत उक्तं कार्ये-
ति । कार्यद्रव्यगतं यन्महत्त्वं तदाश्रयो महद्द्रव्यं । एवं
तत्कारणद्रव्यगतं यदणुत्वं तदाश्रयो ऽणुद्रव्यमेव इति
द्विविधं सिद्धं । परमाणुनिरूपणं चाग्रे भविष्यति ।
इदानीं कार्यरूपद्रव्यस्याश्रयभेदेन द्विविधत्वापनाय
सूत्रम् ॥

कार्यरूपद्रव्यं द्विविधं भौमादिव्यभेदात् ॥ १४ ॥

एतद्विभागानुसारेण ज्योतिःशास्त्रस्य विषय उच्यते ॥

सूर्यादीनां दिव्यपदार्थानां भूपिण्डेन सह सम्ब-
न्धो ज्योतिःशास्त्रस्य विषयः ॥ १५ ॥

अथैते सूर्यादयो दिव्यपदार्थाः कतिविधा इत्याकां-

न्यायां इंग्लंडीयमतानुसारेण ज्योतिःशास्त्रस्य कतिचि
त्सिद्धांतविषयाः संक्षेपत उच्यंते ।   तचादौ उक्तपदा-
र्थानां द्वैविध्यज्ञापनाय सूत्रम् ॥

केषांचित् स्वतःप्रकाशकत्वात् केषांचित् पर-
तःप्रकाशकत्वात् ते द्विविधाः ॥ १६ ॥

ते दिव्यपदार्था द्विविधा इत्यर्थः । कथं तेषां द्वैविध्य-
मित्याह केषांचिदिति । यथा दिव्यपदार्था नक्षत्राणि
सूर्यवत् स्वतःप्रकाशकानि ग्रहाणां चन्द्रवत् सूर्यतेजसा
प्रकाशकत्वात् ते परतःप्रकाशका इति तेषां द्वैविध्यं ॥

ननु चंद्रगतह्रासवृद्धिप्रतीतिश्चक्षुषा भवति शुक्रादी-
नां ग्रहाणां तादृशप्रतीतिः कथं न भवतीति चेदुच्यते ।
ग्रहाणां तादृशप्रतीतिर्न नेत्रमात्रेण भवति किंतु तच
दूरदर्शकयंत्रमपेक्ष्यते ।   अतस्तत्सहकारेणैव शुक्रादिग्र-
हाणां स्पष्टरूपेण ह्रासवृद्धिप्रतीतिर्भवति ॥

अथ भूम्याकारज्ञापनार्थं सूत्रम् ॥

भूगोलाकृतिः ॥ १७ ॥

प्रथमाध्याये न्यायप्रकरणे एतत्साधितं । प्रकारांतरेण साधनंत्वग्रे भविष्यति । एवं भुवो गोलाकृतित्वं भास्क- राचार्येण यथा साधितं तथा प्रसिद्धमेव ॥

अथ भूभ्रमणसिद्धांतसूचनम् ॥

अयं भूगोलः स्वाचमभितो भ्रमन्नहोराचमु- त्पादयति रविं परितो भ्रमंश्च ऋतुभेदानु- त्पादयति ॥ १८ ॥

नन्वेवं सति कस्मात् सूर्यो भूमिं परितो भ्रमतीति प्रतीतिर्लोकानां भवतीति चेदुच्यते । गच्छन्नौस्थबाल- कानां नदीतीरं गच्छतीति प्रतीतिर्यथा भ्रमरूपा तथा उक्तप्रतीतिरपीति बोध्यं । तथाचोक्तमार्यभट्टेन । अनु- लोमगतिर्नौस्थः पश्यत्यचलं विलोमगं यद्वत् अचला- नि भानि तद्वत्समपश्चिमगानि लङ्कायामिति । भूभ्रम- णप्रमाणान्यग्रे वक्ष्यन्ते ॥ अथ चंद्रभ्रमणसिद्धांतसूचनम् ।

चन्द्रो भूगोलं परितो भ्रमति ॥ १८ ॥

___

स्पष्टम् ॥

अथ कथं ग्रहणं भवतीति कदावा भवतीति चा-
पनार्थं सूचम् ॥

___

पृथ्वीमभितो भ्रमंश्चन्द्रो यदा सूर्यपृथिव्योर्म्मध्य
आयाति तदा सूर्यग्रहणं यदाच रविचन्द्र-
योर्म्मध्ये भूरायाति तदा चन्द्रग्रहणं ॥ २० ॥

___

अथ प्रतिमासं पृथ्वीमभितश्चन्द्रस्य भ्रमणेन प्रतिमा-
सं ग्रहणं कुतो न भवतीत्यग्रे विवेचयिष्यते ॥ अथे-
दानीं ग्रहभ्रमणज्ञापनाय सूचम् ॥

___

ग्रहाः सूर्यं परितो भ्रमन्ति ॥ २१ ॥

___

नन्वेवं सति उक्तभ्रमणस्य चाक्षुषप्रत्यचं कुतो न
भवतीति चेदुच्यते । यत द्वयं भूः सूर्यं परितो भ्रमति
अतो भूस्थेन जनेन वास्तविकी ग्रहकचा द्रष्टुं न शक्यते
किंतु गणितेन निश्चीयते ॥

अथ यथा भूश्चन्द्रवती तथा सर्वे ग्रहाः कुतो न चन्द्र-
वन्त इति चेदुच्यते ये ग्रहाः पृथिवीमपेक्ष्य सूर्यस्य समीपे
वर्त्तन्ते तेषां चन्द्रा एव न सन्ति परंतु ये ग्रहा बहुदूरं
वर्त्तन्ते तेषामनेके चन्द्राः सन्तीति ते चन्द्रवंत एव ॥

अथ केषां ग्रहाणां कति चन्द्राः सन्तीति ज्ञापनार्थं
सूत्रम् ॥

बृहस्पतेश्चत्वारश्चन्द्राः शनेः सप्त अ-
स्य परितो महाचक्रमपि
वर्त्तते ॥ २२ ॥

शनिचक्रं दूरदर्शकयन्त्रमन्तरेण स्पष्टं न दृष्यते ।
एवंच चक्षुर्मात्रेण यावन्ति नक्षत्राणि दृष्यन्ते तेभ्यो
ऽन्यानि कोटिशो दूरदर्शकयन्त्रेण दृष्यन्ते । अथ मंदा-
किन्यपि दिव्यपदार्थः अतस्तत्स्वरूपज्ञापनार्थं सूत्रम् ॥

मन्दाकिनी नक्षत्रपुञ्ज एव ॥ २३ ॥

खेया मन्दाकिनी सा नक्षत्रपुञ्ज एवेति दूरदर्शकय-
च

न्वेषण ज्ञायते । अन्ये ज्योति:सिद्धांता: श्री बापूदे-
वरचितखगोलवर्णने प्रकाशिता:   ॥

अथैतदध्यायस्य चतुर्दशसूचोक्तविषयानुसारेण भूपि-
ण्डस्थितानां कार्याणां किंचिद्विस्तरेण वर्णनं क्रियते   ॥

भूपृष्ठस्थितकार्याणां व्यवस्था तत्कार्यस्वरूपाणि
तत्कार्यव्यवस्थाया: कारणानिचेति त्रयो
भूपिण्डस्थनिरूपणविषया: ॥ २४ ॥

अथ भूपृष्ठस्थितकार्याणां व्यवस्था का इत्याकां-
क्षायां सूत्रम् ॥

भूपृष्ठस्थितकार्याणां व्यवस्था
भूपृष्ठविद्या ॥ २५ ॥

भूपृष्ठस्थितानां समुद्रनदनदीपुरग्रामपर्वतादिदे-
शानां परस्परसम्बन्धस्य व्यवस्था भूपृष्ठविद्या इ-
त्यर्थ: ॥

अथ नगरस्थित्यनुसारेण नद्यो न वहन्ति परंतु प्राय-
शो नदीजललाभार्थं तत्तीरे महानगराणि निर्मिता-
नि भवन्ति यथा गङ्गातीरे काशी । किंतु पर्वतव्यवस्थि-
त्यनुसारेण नद्यो वहन्त्येव इत्येतस्य किं कारणमिति
जिज्ञासायां तज्ज्ञापनाय सूचम् ॥

पर्वताधीननदीप्रवाहात् पर्वतस्थित्य
नुसारेण नद्यो वहन्ति ॥ २६ ॥

पर्वतशब्देन सर्वेषामुच्चभूम्यंशानां ग्रहणं । हिमा-
लयादिषु महापर्वतेषु गङ्गादीनां महानदीनामुत्प-
त्तेर्नदीप्रवाहाणां पर्वताधीनत्वं अतः पर्वतस्थित्यनु-
सारेणैव नद्यो वहन्तीति ज्ञेयं । अथ महापर्वते कथं
महानद्युत्पत्तिरिति चेत् तज्ज्ञापनार्थं सूचम् ॥

महापर्वतशिखराणि शीतानि तच काचित् सीमा
वर्त्तते यत ऊर्ध्वं वर्तमानानि पर्वतशिखरा-
णि सर्वदा हिमाढतानि भवन्ति ॥ २७ ॥

सामपि तद्विकार: स्यात् तासु चाल्पवेगवत्योऽधिकष्ट–
द्विमत्य: स्यु: किन्तु नहि बह्व्योऽन्या उदग्वाहिन्यो नद्यो
नीलनदीवद्विकृता भवन्ति ।   अथ द्वितीयवर्णनम् ।
वर्णयितुमते समुद्रो ऽपि नदी भवति तेनेयं पृथ्वी आट–
तास्ति ।   अतोऽस्या नद्या: समुद्रोत्पन्नत्वादियं वर्त्तते
इति परमेतदपिवर्णनमसमीचीनम् ।   यत: समु–
द्रविषये यदत्र वर्णितं तत्तु प्रमाणशून्या अस्पष्टार्था
किंवदन्ती अस्ति ।   समुद्रोऽपि नदी भवतीति तु न
मया श्रुतम् ।   यत्तु कदाचित् होमराख्योऽन्ये वा
कतिचन प्राचीनकवयस्ताट्टशमर्थं लब्ध्वा तं स्वकाव्यग्र–
न्थेषु वर्णयाञ्चक्रुस्तत्तु चमत्कारार्थमेव ।   तृतीयवर्ण–
नम् ।   हिमद्रवीभवनेन नदी वर्त्तते इति ।   नैतदपि
सुन्दरम् ।   यत: इयं नदी याम्यदेशान्निर्गता सौम्य–
दिशं वहन्ती सती मिश्राख्यदेशमवरोहति ।   एवं
उष्णदेशाच्छीतदेशमायान्त्या एतन्नद्या इद्धि: हिमा–
ज्जायत इति कथं वक्तुं शक्यते ।   हिमजातट्ड्यनुप–
पत्तौ बहवो हेतवो भवन्ति येभ्यो यस्य कस्यापि तद्वि–

षये संगीतिविच्छित्तिः स्यात् । तत्रादौ । नद्युत्प-
त्तिदेशादुष्णो वायुर्वहतीति उक्तट्ड्ग्यनुपपत्तौ स्फुटो
हेतुः । अथ च तद्देशीया मनुष्या उष्णतया कृष्ण-
वर्णा भवन्ति किञ्च उष्णदेशप्रिया आकाख्यः सदैव तत्र
वसन्ति क्रौञ्चपच्छिणाश्च शीतकाले सौम्यदेशादागत्य तत्र
वसन्ति । परं यदि तत्र किञ्चिदपि हिमं वर्षेत् तदेदं
सर्वं न घटेत ॥

सकलयाऽनया हीरादत्तस्योक्त्या स उच्चपर्वतस्थो
ष्णताया उष्णदेशस्थोष्णतायाश्च भेदं नावगच्छेदिति
स्फुटमवगम्यते । स भेदोऽस्य ग्रन्थस्य द्वितीयाध्याये
सप्तविंशसूत्रे वर्णितोऽस्ति ॥

अथ ये पर्वतसम्बधिघटना जानन्ति ते नीलाख्यनदी
ग्रीष्मकाले वर्द्धत इत्यस्य अवगममनु इयं नदी हिम-
श्रेण्या उत्पद्यत इति तर्कयेयुः ॥

अथ यथा गङ्गया हिमस्थानान्निर्गमप्रत्यक्षः तथा-
ऽधुना नीलनद्या अपि प्रत्यक्षोऽभूत् ॥

अथ यो हि हिमालयपर्वते परिभ्रम्य अगाण्येकू-

टान् तदुत्पन्नप्रवाहाँश्च व्यक्तानीच्चेत स तेषु कश्चन
क्रमसंस्थितिर्वर्तत इति कदाचिदपि नाहेत किन्तु यो
हि भङ्ग्विद्वारा तान् युगपत् पश्येत् स तेषु कश्चन नि-
यतः क्रमः संस्थितिश्चास्तीति जानीयात् । संचि
त्सवर्णनस्याकांच्चायां तच सूचम् ॥

—————

भारतवर्षस्योत्तरादिग्भि पूर्वापरो हि-
मालयपर्वतस्य विस्तरः ॥ २९ ॥

—————

अस्मिन् पर्वते अनेके मार्गाः सन्ति यानारुह्य या-
चिण : पारेपर्वतं गन्तुं श्कुवन्ति । ते घट्टाख्या मार्गा
यस्यां हिमाद्रितपर्वतश्रेण्यां वर्त्तन्ते सा पर्वतश्रेणी हिम-
घट्टश्रेणीत्यभिधीयते ॥

अथ मुख्यशिखराणामपेच्चायां सूचम् ॥

—————

हिमालयपर्वतस्य मुख्यशिखराणि स्व-
स्वपर्वतश्रेणीसहितानि घट्टश्रेणीद-
च्चिणादिग्भि विस्तृतानि सन्ति ॥ ३० ॥

—————

गङ्गानद्युत्पत्तिस्थानं यद् जम्नौचीशिखरं तत्प्रभृति
यथाक्रमं मुख्यशिखरनामानि कथ्यन्ते । तथाहि ।
जम्नौची नन्ददेवी धवलगिरि: गोस्वामिस्थानं कर्षं-
गिरि: चमलारी गिरियुग्मञ्च ॥

अथ तत्सम्बन्धिनदीव्यवस्थासूचम् ॥

_____

तच्चत्पर्वतश्रेढ्यन्तर्गता बहव: प्र-
वाहा: क्रमेण एकीभूय तच्च तच्च एकां
नदीमुत्पादयन्ति ॥ ३१ ॥

_____

तथाहि जम्नौचीपर्वतो ऽस्ति यस्योच्चता १७११२ ह-
स्तपरिमिता अपिच नन्ददेवी यस्योच्चता १७१२८ ह-
स्तपरिमिता । तच्छिखरद्वयसम्बन्धिपर्वतश्रेणीद्वय-
मध्ये ये ये गलितहिमोत्पन्ना: प्रवाहा: सन्ति ते क्रमेण
मिथ: संयुक्ता भवन्ति तःस्थानस्य कटाहाकृतित्वात् ।
अतस्तच्च अनेकेषु अनल्पेषु पर्वतेषु गङ्गाचीतीर्थरूमी-
पस्थकेदारनाथाख्यादिषु सत्त्वपि सर्वे सरित एकीभूय
गङ्गानदी भवन्ति ॥

छ

एवमेव नन्ददेवीशिखरोपश्रेण्या धवलगिरिशिख-
रोपश्रेण्याश्च मध्ये अनेका नद्यः एकीभूय घर्घटसन्निकां
कर्णाटिनदीमुत्पादयन्ति । सा च गङ्गया सह स-
ङ्गमेति ॥

पुनः धवलगिरिशिखरगोस्वामिस्थानशिखरयोर्मध्ये
सप्तगण्डक्याख्याः प्रवाहा वहन्ति ते च मिलित्वा गण्ड-
कनदीति संज्ञां प्राप्य गङ्गया सङ्गता भवन्ति ॥

पुनरपि गोस्वामिस्थानशिखरकञ्चंगिरिशिखरयोर्म-
ध्ये सप्तकौशिकाख्या नद्यो वहन्ति तास्त्वैकीभूय कौशीति
आख्यां लब्ध्वा गङ्गया सह मिलन्ति । एवमन्येष्वपि
ताट्टग्येषु स्थानेषु भवतीति बोध्यम् ॥

अथ महापर्वतप्रदेशं त्यक्ता ता नद्यो गङ्गासङ्गात्
प्राक् कीट्ग्ये प्रदेशे वहन्तीत्याकांच्चायामाह ॥

एता नद्यो हिमाचलसमान्तरदेशे व-
र्त्तमानां सैकतपाषाणपर्वतश्रेणीं भि-
त्त्वा स्वस्वसरणिं कुर्वन्ति ॥ ३२ ॥

इयं सैकतपाषाणपर्वतश्रेणी उभयतो ऽपि भूपृष्ठात्
द्विशतीहस्तप्रभृतिचतु:शतीहस्तपर्य्यन्तोच्चा भवति ।
पुन: सूचयति ॥

सैकतपाषाणपर्वतश्रेण्या उत्तरभागे धू-
नाख्यप्रदेशाद्दक्षिणादिदिशि भावराख्यस्तदध:-
तराईसंज्ञकश्चेति प्रदेशा भवन्ति ॥ ३३ ॥

हिमालयविस्तृतिसममितिरेकोनचत्वारिंशत्क्रोशा भ-
वन्ति । तत्र प्रत्येकं चयोदशक्रोशपरिमाणास्त्रयो वि-
भागा: स्वदेशविशेषानुसारत: क्रमेणाधरमध्योर्ध्वसं-
ज्ञा: स्यु: । तच्चाधरविभागे तराईसंज्ञको भावरसंज्ञ-
कश्च सैकतपाषाणपर्वतश्रेणीच तत्सम्बन्धिनो धूना-
ख्यप्रदेशाश्च वर्त्तन्ते । अयं विभागो भूपृष्ठप्रभृति: स-
मुद्रपृष्ठात् २७०० हस्तोच्छायो य: प्रदेशस्तदवधिर्भव-
ति । तत: समुद्रपृष्ठात् ६७०० हस्तोच्छायो य: प्रदे-
शस्तदवधिर्मध्यविभागो भवति । ऊर्ध्वविभाग: शेष: ॥

अथ पर्वतशिखरप्रदेशाः शीतला भवन्तीति सप्त-
विंशसूचे प्रोक्तं । तेन यदि शिखरदेशे हिमं स्यात्
तलदेशेचात्युष्णो वायुः स्यात्तर्हि तलाच्छिखरपर्यन्तं
प्रदेशे उत्तरोत्तरमवश्यं शीतलो भवेदिति स्पष्टतरं ।
अथ कतिचन जन्तव उष्णदेशप्रिया अन्येच शीतप्रिया
भवन्ति । एवमोषध्यो ऽपि । तथा जन्तव ओषधयश्च
स्वस्वप्रियप्रदेशे पुष्टिमायान्तीति सर्वस्मिन् जगति स्थि-
तिरस्तीत्यवगतमपि तत्सर्व हिमालयप्रदेशे दृष्टा-
न्तितं । तच्च सूचयति ॥

हिमालये ऽन्येषुचोच्चपर्वतेषु प्रदेशोच्चता-
विशेषे ओषधीनां जन्तूनाञ्च विशेषा
भवन्ति ॥ ३४ ॥

तथाहि । ओषध्यः । अधरविभागे साल: शिंशपा
तुन्द: पलाशो वट: पिप्पलश्चेत्यादयो भवन्ति । म-
ध्यविभागेच पूर्वोक्तविलक्षणा इङ्गलण्डीयादिदेशेषु प्र-
सिद्धा वृक्षा भवन्ति येचोष्णदेशे नोत्पद्यन्ते । अथ इङ्-

लक्षद्वीपादिदेशानामप्युत्तरतो ३तिशीतदेशेषु प्रसिद्धा
ये हच्चास्तत्सजातीया देवदारुभूर्ज्जादयो हच्चास्तदू-
र्ध्वभागे उत्पद्यन्ते ॥

जन्तवस्तु अधरदेशे उष्णादेशप्रिया वसन्ति किन्तु म-
ध्यदेशे ताट्टशाः प्राणिनो वर्त्तन्ते ये ३धरदेशे उष्णता-
धिक्यात् स्थातुं न शक्नुवन्ति । अथाधरमध्यदेशयो-
र्पि उष्णताबाहुल्यात् स्थातुमसमर्था जन्तव ऊर्ध्वदेशे
वसन्ति । एवमेव अधरप्रदेशे हस्तिनः खङ्गा व्याघ्रा
हरिणाश्वेत्यादयो वसन्ति । मध्यदेशे न हस्तिनो
न खङ्गा नवा व्याघ्राः । तच्च केवलमेका हरिण-
जातिर्वसति । ऊर्ध्वदेशेतु नैतेषां पूर्वोक्तानां ह-
स्त्यादीनां कश्चन जन्तुरास्ते किन्तु तच्च पर्वतीयमेषा
पर्वतीयाजाश्वेत्यादयस्तिष्ठन्ति । एवमधरप्रदेशजाः
काका नोर्ध्वदेशे दृश्यन्ते । ताट्टशाः कतिपये म-
ध्यदेशे वसन्ति ॥

एवं हिमालये तलाच्छिखरपर्यन्तमुत्तरोत्तरं शैत्य-
वृद्धेः तादृशा देशविशेषा भवन्ति ये प्रायः समानभूम्यां
सुमहति देशे भवन्ति ॥

अथ युरोपीयाख्यैः कल्पितानां भूपृष्ठविभागानां
वर्णनम् ॥

एशिया आफ्रिका यूरोपा अमेरिकाख्या-
स्त्वारो भूपृष्ठखण्डा ज्ञेयाः ॥ ३५ ॥

भूपृष्ठखण्डा इति । पृथ्वी गोलप्राया भवति यस्या
व्यासः किञ्चित्सान्तरश्चतुस्सहस्रक्रोशमितो भवति ।
भूपृष्ठस्य किञ्चिदधिकौ द्वौ तृतीयांशौ समुद्रेण व्याप्तौ
भवतः । शेषः स्थलम्भवति । तस्यैव चत्वारः खण्डा
इत्यर्थः । अथ तल्लक्षणानि ॥

भारतवर्षं यत्र वर्त्तते स एव एशियाख्यखण्डः
इङ्ग्लण्डसंज्ञदेशो यत्र स एव यूरोपाख्यो
ऽपिच एशियाख्यखण्डस्य प्रतीच्यां दिशि आ-

फ्रिकायूरोपाखण्डे वर्त्तेते आफ्रिकाखण्डस्य-
चोदीच्यां दिशि यूरोपाखण्डं वर्त्तते भारत-
वर्षाक्कुदलान्तरे अमेरिकाख्यखण्डन्तिष्ठति
इदमुत्तरध्रुवसमीपप्रदेशादि दचिणध्रुवा-
न्तिकप्रदेशपर्यन्तं विस्तृतं भवति ॥ ३६ ॥

सर्वे जलस्थलस्थाः प्रदेशा भूपृष्ठभङ्गिप्रदेर्शनेन च्या-
यन्ते । अथ कतिचिद्देशा उष्णा भवन्ति कतिचना-
नुष्णा इत्यच मुख्यो हेतुरुच्यते ॥

र्विकिरणलम्बरूपतिर्यक्पतनानुसा-
रतत्तद्देशस्योष्णता ॥ ३७ ॥

यो देशो निरच्चासन्नो भवति तच रविकिरणा ल-
म्बरूपाः पतन्ति अतः स भारतवर्षवदुष्णो देशो भ-
वति । यश्च देशो ध्रुवासन्नो भवति तच रविकिर-
णास्तिर्यक् पतन्तीति स देशः शीतो भवति यथा इङ्ग-
लण्डाख्यः ॥

पर्वतेभ्यः सरित उत्पद्यन्ते यथा हिमालयाङ्गङ्गाद्य उत्पन्ना इति पूर्वमुक्तं । आफ्रिकादेशे अल्पाः पर्वता इति तच्च सरितो अल्पा अतः स देशो निष्फलप्रायः ॥

भूगोलग्रन्थेषु विषयाणां ग्रामाणाञ्च दिशो लिख्यन्ते तन्तद्देशसम्बन्धीनि नीतिधर्माचरणानि राज्यप्रकरणानिच प्रदर्श्यन्ते । तान्यच्च विस्तरभयान्न लिख्यन्ते ॥

इति भूपृष्ठव्यवस्थाप्रकरणं समाप्तम् ॥

अथ भूपृष्ठव्यवस्थां संक्षेपेणाभिधाय तच्चत्यभौतिक-कार्याणि तदीयकारणानिच निरूप्यन्ते । भौतिककार्यसामान्ये कारणानि परमाणवः । तेषां विशेषधर्मच्चयनिरूपणाय सूचम् ॥

अथ परमाणुतदाकर्षणोत्सार-णजडत्वजिज्ञासा ॥ ३८ ॥

परमाखिति । भौतिककार्यमाचमतिसूच्च्मांशैर्भे-

द्यमस्ति तेषांश्च मनुष्यशक्तियेन नाश्था न भवन्ति ।
यथा । लोहादिधात्वंश: सहस्रवारं चुम्बो भग्नश्छि-
न्नो गलितो विकृतो ऽपि सदा यथावत् दर्शयितुं श-
क्य: । तत्र ये सूक्ष्मतमा अंशा अभेद्या अछेद्याश्च भ-
वन्ति ते परमाणुसंज्ञका: स्यु: ॥

तदाकर्षणेति । परमाणव: पृथक्भूत्वा कार्यीभूय
वा अन्यपरमाणून् प्रति जिगमिषवो भवन्तीति दृ-
श्यते । यथा पाषाणादिकार्यस्यांशा: केनचिद्बलेन
संयुक्तास्तिष्ठन्तीति दृश्यते । यथा वा पतित: पाषाण:
केनचिद्बलेन भूम्या सह संयुक्तास्तिष्ठतीति दृश्यते ।
यथा वा समुद्रश्चन्द्रम्प्रति जिगमिषुर्दृश्यते । अत्र का-
रणं तत् आकर्षणाख्यं स्यात् ॥

उत्सारणेति । परमाणूनां परस्पराकर्षणे क्वचित्प्र-
तिबन्धो दृश्यते । यथा यत्रोष्णताया: प्रवेशस्तत्र । तदा
परमाणूनां मिथो विश्लेषो लक्ष्यते । तथाहि । हि-
ममुष्णता प्रविशेत्तर्हि जलमुत्पद्यते । अपिच जल-
ज

मुष्णता प्रविशेत्तर्हि वाष्प उत्पद्यते । अत्र कारण-
मुत्सारणाख्यं स्यात् ॥

जडत्वमिति । यथा कुम्भकारस्य चक्रं भ्रमणकाले
प्रथमं भ्रामकयत्नं रुणद्धि किन्त्वनन्तरं क्रमभः तद्य-
लभङ्क्यनुसारेण शीघ्रगतिं लभते तथाच स्वभ्रमानु-
सारेण स्वगमनप्रतिरोधयत्नं प्रतिकरोति तथैव सर्वाणि
कार्याणि परमाणवश्च स्थैर्यगमनयोरपेच्चायां केनचि-
द्विकारेत्सुल्बेन विशिष्टा इति दृश्यते । तच कार्यं
जडत्वसंज्ञं स्यात् ॥

अथ परमाण्वाकर्षणादिधर्मविचारात् इन्द्रियग्राह्य-
गुणवत्त्वंसारधर्मत्वानसम्भवो यथाशक्ति प्रदर्श्यते ॥

इन्द्रियग्राह्यगुणवत्त्वंसारो ऽयं
परमाणुनिर्मितः ॥ ३८ ॥

लोहादिधात्वंशः सहस्रवारं चूर्णो भग्नो गलितो
विकृतोऽपि सदा यथावत् दर्शयितुं शक्य इति पूर्वं

कथितम् । यद्यपि दृच्चपचादीनां पच्चिपच्चादीनां वा
तादृशो धर्मो नास्ति यतो जीविल्वसाधनविशिष्टपदा-
र्थकर्तृत्वमखदादीनामसम्भावि तथापि दृच्चादीनां ज्व-
लनादिकाले ऽप्येको ऽपि परमाणुर्न नश्यतीति रसा-
यनप्रकरणे प्रदर्श्यते ॥

अथ परमाणुपरिमाणमुच्यते ॥

परमाणवो ऽति सूच्ष्माः ॥ ४० ॥

तच्चैकं प्रमाणमुच्यते तद्यथा सुवर्णकारास्ताडनेन
सुवर्णपचाणि तथा सूच्ष्मीकुर्वन्ति यथा ३६००००० संख्या-
कानां तेषां पचाणां खोपर्युपरि न्यस्तानां पिण्ड र-
काङ्गुलमाचपरिमितो भवति । तथापि प्रत्येकं पचे प-
रमाणुपरिमाणादधिकघनत्वमवगम्यते यतो राजततारे
लिप्तस्य स्वर्णस्य घनत्वं तादृशपचघनत्वाद्न्यूनं लक्ष्यते ।
तस्यापि घनत्वस्य परमाणुपरिमितत्वे मानाभावः ॥
परमाणूनाम्परिमाणं सर्वपरिमाणेभ्यः सूच्ष्मं भवति

तच्च न ह्यणुकपरिमाणोत्पत्तौ कारणं अणुजन्यस्या-
णुतरत्वप्रसङ्गात् किन्तु तच्चान्यदेवेश्वरेच्छादिकमगत्या
कारणं वक्तव्यमिति न्यायसिद्धान्तः । अस्मत्सिद्धान्त-
स्तु तत्तज्जातीयपरमाणूनां तत्तज्जातीयसर्वद्रव्येभ्यः
सूक्ष्मतर्त्वे ऽपि तत्तज्जातिनियतमहत्त्वस्वीकारे न कि-
ञ्चिदपि बाधकं । एवञ्च परमाणुपरिमाणं ह्यणुक-
परिमाणे एकं कारणमित्यस्योपपत्तिरग्रे वक्ष्यते ॥

यत्तु निष्परिमाणा एव परमाणवो विरोधनामके
वक्ष्यमाणधर्मश्च तेषां स्वाभाविक एवेति यूरोपदेशीयस्य
बास्कविकाख्यस्य मतं । साक्षाद्बाधविरहे ऽपि भौ-
तिकवस्तुतत्त्वनिर्णये ऽल्पतरोपकारकत्वात् अस्मन्मता-
पेच्चया न वरम् । अत उच्यते ॥

भौतिकपदार्थाः परिमाणरहिताः
कदापि न सन्ति ॥ ४१ ॥

यथा तूलराशिः सङ्कोचकव्यापारेणाल्पतरपरिमाणः

कर्त्तुं शक्यते नतु निष्परिमाणः तथा सर्व एव भौ-
तिकपदार्थाः परिमाणं कथमपि न त्यजन्ति । स एष
स्वीयपरिमाणापरित्यागलक्षणो धर्मो विरोध इत्युच्य-
ते । तत्स्वरूपञ्चैकदैकदा द्वयोर्बहूनां वा द्रव्याणां
स्थित्यभावः । तच्चोदाहरणान्युच्यन्ते ॥

जलेन परिपूर्णे पात्रे पाषाणः प्रक्षिप्यते चेत् पाषा-
णाय अवकाशं दत्त्वा जलं वहिर्गच्छति ॥

एका काचनलिका अङ्गुलीप्रतिरुद्धोपरिमुखी अना-
च्छादितार्धोमुखीच जले निमज्जिता सती जलेन न पूर्यते
तदन्तर्गतवायुना जलस्य विरोधात् । यदाच वायु-
निर्गमनाय अङ्गुलिनिष्कासनं क्रियते तदा वहिःस्थित-
जलपृष्ठपर्यन्तं जलपूर्णा भवति ॥

वायुपूर्णे ऽधोमुखे पात्रे जले स्थापिते किञ्चिज्जल-
मन्तर्गच्छति परन्तु यावत् तत्पात्रमुखमुपरि न कृत्वा
वायोर्गमनमार्गो न कृतस्तावत् तत्पात्रं जलेन पूर्यन्न
भवति वायुविरोधात् । अन्तर्वर्त्तमानदीपे ऽधोमुखे

अनाकुलास्तिर्यगधः स्थिताश्च
तिष्ठन्ति ते तच्च वयं यथाच ।

परमाणूनां एककेन्द्रगामित्वादेव पृथ्वी गोलरूपा
जाता । एवं संगताः कुज्झटिकांश्च अपि गोलरूपा
दृश्यन्त । तेषुच भूमिम्प्राप्तेषु तेषां गोलत्वं सुतरां
नश्यति परन्तु द्रवीकृतसीसके उच्चस्थानाच्चालनीद्वारा
पातिते तद्बिन्दवो घनीभूय भूमिं प्राप्ता गोलत्वं न त्य-
जन्ति । अनयैव युक्त्या व्याधा गोलरूपान् सूच्चमसी-
सककणान् पच्चिछननार्थं निर्मायन्ते । अद्रवीकृतसी-
सकांश्च उच्चस्थानात्पातिता अपि गोलत्वं न प्राप्नवन्ति
अत एव पृथ्वी चन्द्रः सूर्यो ग्रहाश्च पूर्वं कस्मिंश्चित्
काले द्रवीभूता आसन् इत्यनुमीयते । अपिच पर-
स्पराकर्षणरूपः सामान्यधर्मो ऽपि तेषु वर्त्तते इत्य-
नुमीयते ॥

ननु यदि सर्वमत्र भूकेन्द्रम्प्रति जिगमिषुः स्यात्तर्हि
धूमः कथमूर्ध्वं गच्छतीति चेदुच्यते । धूमः स्वय-

मवोर्ध्वं न गच्छति किन्तु भूसमीपस्थो वायुस्तस्माद्गुरु-
तरो ऽस्तीति सो ऽधो गत्वा धूममुत्थापयति यथा
घटस्थजले तले मुक्तं तृणं ततो गुरु जलमुत्थापयति ॥

परस्पराकर्षणरूपो धर्मो ऽतिदूरप्रदेशे ऽपि न न-
श्यति । तचोदाहरणं । चन्द्रो यद्यपि १०५६००
एतत्क्रोशमितदूरदेशे वर्त्तते तथापि स्वकीयाकर्षणेन
स्वाधःस्थितसमुद्रस्य नीरमुत्थापयति । एवं समुद्रस्य
वृद्धिह्रासौ जायेते । तथैवातिदूरे वर्तमानो ऽपि सूर्यः
समुद्रमाकर्षति । यदा पृथ्वी सूर्यः चन्द्रश्च एकस्यां
सरलरेखायामायान्ति तदा सूर्यचन्द्रयोराकर्षणैक्या-
त्समुद्रो ऽधिकं वर्धते ॥

अथ यद्यप्यतिदूरप्रदेशे ऽपि आकर्षणरूपो धर्मो न
नश्यति तथापि दूरत्वेनाल्पीक्रियत इति सूचयति ॥

भौतिकपदार्थानां परस्परसन्निधानतारत-
म्यात् परस्पराकर्षणतारतम्यं भवति यथा

दीपसामीप्यदूरत्वविषघान्तत्तेजस
आधिक्यन्यूनत्वविशेष: ॥ ४३ ॥

हस्तपरिमितवर्गाकारकाष्ठखण्डं स्थापयित्वा अर्द्धह-
स्तदूरस्थेन दीपेन प्रकाशयेत् तदा तत्काष्ठखण्डस्य
पश्चात् द्वितीयं द्विहस्तपरिमितकाष्ठखण्डं स्थापयेन्त-
र्हि प्रथमखण्डस्य च्छायया द्वितीयखण्ड आच्छादित
एव भविष्यति । अथ द्विहस्तपरिमितभुजवर्गे एक-
हस्तभुजपरिमिता: चत्वारो वर्गा भवन्ति यतो नह्यच
भुजस्य केवलं दैर्घ्यं वर्द्धते येन वर्गो द्विगुण: स्यात् किन्तु
विस्तारो ऽपि तथैव हृद्विमेति अत: स चतुर्गुणो
भवति । अतस्तेज: स्वनिष्काशनस्थानाद्द्विगुणे ऽन्तरे
चतुर्गुणप्रदेशे विस्तृतत्वात्तस्य तेजस्त्वं स्वचतुर्थांशसमं
भवति । एवमनया युक्त्या त्रिगुणे ऽन्तरे तत्त्वनवमां-
शतुल्यं जायते । एवमग्रे ऽपि । एवं तेजसो विस्तृति-
विशेषात्तेजस्त्वह्राससविशेष: तथा भवति यथा कति-

पर्यैर्मनुष्यैः समं विभज्य ग्राह्यासु कतिषुचिन्मुद्रासु यथा
मनुष्यसंख्याधिक्यविशेषात्प्रतिमनुष्यं समं तन्मुद्राभाग-
न्यूनताविशेषः । वस्तुतः प्रकाशौष्ण्याकृष्टिध्वन्यादयः
कदम्बगोलकन्यायेनोत्पद्यमानाः सर्व एव गुणाः पू-
र्वोक्तान्योन्यपिदधद्वर्गेच्चेत्पट्ह्यानुसारतो स्वीयभावह्रासं
यान्ति । एवमेवोक्तनियमानुसारेण यस्य वस्तुनो भारः
समुद्रतीरे सहस्रपलप्रमाणस्तस्यैव भारः किञ्चिदुच्चप्र-
देशे शैले पञ्चभिः पलैर्न्यूनो भवति । इदं स्थिति-
स्थापकविशिष्टतुलया परीच्य साधितमस्ति । तादृ-
ग्वस्तुलाया उच्चपर्वते यावद्दैर्घ्यं तदपेच्चया समुद्रतीरे
ऽधिकं भवतीत्येतेन तच्च भाराधिक्यं द्योत्यते । अथो-
च्चपर्वते यो भारह्रासो भवति तत्प्रमाणानुसारेण गणि-
ते कृते वस्तुनः सहस्रपलप्रमाणो भारः भुवं प्रति प्रा-
वर्ग्यं वा चन्द्रकच्चायां विंशतिमाषप्रमाणं सिध्यति ।
एतद्दर्शनमुत्तरभागे पुनः करिष्यते ॥

यदा घटाज्जलं पात्यते तदा घटकृताकृष्टिहेतोस्त-

जलं नहि तत्क्षणादेव तस्मात्पृथग्भवति । किन्तु त-
दाकर्षकभाण्डबहिर्देशे संलग्नं भूत्वा पतति। अत एव
जले यद्भाण्डकृताकर्षणं तन्निरसनाय नलीविशिष्टानि
पात्रविशेषाणि निर्मीयन्ते । ताद्दृशानि पात्राणि कमण्ड
ल्वाख्यानि प्रायो भारतवर्षीयतपस्विनां निकटे दृश्यन्ते ।
अथ यदि नलीरहितपात्रात्तथा जलं पातयितुमिष्यते
यथा तत्पात्रबहिर्देशे ऽलग्नं सत् पतेत् तदा पात्रमुखे
जलपतनस्थले काचादिदण्डं संस्थाप्य पातयेत् तथाच
तत्पात्रकृताकृष्टिविरुद्धं दण्डकृताकर्षणं भवति तेन
दण्डद्वारा सर्वं जलं पात्रालग्नं सत् बहि: पतति ॥

जलकणा मिथ: संलग्ना भवन्ति यथा सूच्ष्मा सूची
अपि जलपृष्ठे मन्दं मोचिता सती तत्र प्लवते । तत्सूची-
भारो जलपृष्ठं भेत्तुं न शक्नोति । एवमेव अल्पा:
कीटमच्चिकादय: पानीयपृष्ठोपरि वार्योपरिस्था गन्तुं
शक्नुवन्ति ॥

अन्योन्यलग्ने जलमग्नाधरभागे द्वे काचादिपत्रे खम-

थ्यदेशे ऽन्योन्यसान्निध्यानुसारतो जलमाकर्षतः । तेन
जलं खपृष्ठादूर्ध्वं गच्छति । एवमेव पानीयस्य मस्यास्तै-
लस्य वा बिन्दुः पुस्तकपार्श्वभागे पतितः सन् शीघ्रं प-
श्चात्तः प्रविश्य दूरं प्रसरति । ऋथच दीपवर्त्तिर्द्वि-
चाङ्गलप्रमाणदूरदेशादपि तैलमाकर्षति ॥

परमाणवः खान्तर्गतोष्णतया विरला भवन्तीत्यत्र
प्राक् तस्मिन् विषये किञ्चिद्विस्तरेण कथनाय सूचम् ॥

————

परमाणवः खान्तर्गताया उत्सारणकर्त्रा उ-
ष्णताया न्यूनत्वानुसारेण सान्द्रीभूय आ-
धिक्यानुसारतो वा विरलीभूय अद्रवा द्रव-
रूपा वायुरूपाश्च जायन्ते ॥ ४४ ॥

————

परं नहीदानीमुष्णतायाः खरूपमुच्यते किन्तु सा
केवलं उत्सारणे हेतुर्भवतीत्युच्यते ॥

भौतिकपदार्थे यथायथोष्णता वर्द्धते तथातथा त-
त्परमाणूनां विरलीभवनात्स पदार्थो विस्तृतो भवति ।

एवं घनपदार्थः पूर्वं विस्तृतो भूत्वा मृदुर्भवति । त-
तो विलीयते । अन्ते तत्परमाणवो ऽतिविरला भ-
वन्ति येन स पदार्थः स्थितिस्थापकविशिष्टः प्रवाही
नाम वायुरूपो जायते । अथैताद्दृशो वायुरूपस्तद-
न्तर्वर्त्युष्णताया निष्क्राश्नेन पुनस्त्रीयावस्थे प्राप्नोति ॥

तदिदृथं । हिमे उष्णीकृते तत्पानीयं जायते
पानीयेचात्युष्णीकृते तद्बाष्पं जायते । अथ पुनर्वाष्पे
शीतीकृते तत्पानीयं भवति । पानीयेच शीतीकृते
तत्पनर्हिमं जायते । एवमेकस्यैव पदार्थस्य हिमत्वं
पानीयत्वं वाष्पत्वञ्चेति त्रयो भावा भवन्ति । एवम-
न्य ऽपि भौतिकपदार्था उष्णतया नानाभावा भवन्ति
किन्तु केचन अधिकोष्णतामपेच्चन्ते केचन न्यूनामिति
विशेषः । अन्यथा अद्रवा द्रवरूपा वायुरूपाश्च ए-
कस्मिन्नेव काले न दृष्येरन् ॥

उष्णतामापनार्थमुष्णतामापकाख्यमेकं यन्त्रं क्रियते ।
तन्निर्माणविधिरग्रे वच्यते ॥

अथाकर्षणोत्सारणयोर्विशेषतो ये तदाधीनधर्मवि-
शेषा भवन्ति तत्कथनार्थं सूचम ॥

आकर्षणोत्सारणयोर्विशेषत: सञ्जाता धर्मविशे-
षा: स्फटिकाकारग्रहणशीलत्वं सान्तरत्वं दृढत्वं
काठिन्यं स्थितिस्थापको भङ्गुरता ताडनवर्द्धनी-
यता प्रसरच्चमता स्नेहाकर्षणञ्चेति ॥ ४५ ॥

स्फटिकाकारग्रहणशीलत्वमिति । यदा यदा लव-
णजातिर्जलसंयोगादिविलीयते तदा तदा उष्णतया तच्-
त्यजले शोषिते सा एकविधमेवाकारं गृह्णाति नान्य-
मिति केषाञ्चित्पदार्थानां सर्वदा स्वीयाकारनियमवि-
शिष्टोत्पत्तिदर्शनादिदमनुमीयते यत् परमाणव: का-
र्यारम्भकाले प्रतिपरमाणून् सर्वावयवावच्छेदेन नाक-
र्षन्ति किन्तु केनचिद्देकेनैवेति । तच्च यो धर्म: स स्फ-
टिकाकारग्रहणशीलत्वसंज्ञ: स्यात् प्राय: स्फटिकविशेषे
तद्धर्मस्य दर्शनात् ॥

सान्तरत्वमिति । कठिनपाषाणस्याप्यवयवाः सान्तरा
भवन्ति ।   अत एव सैकतपाषाणविशेषस्य पानीयशो-
धने उपकर्तृत्वम् । तच्च कारणं सान्तरत्वसंज्ञं स्यात् ॥

दृढत्वमिति ।   घनहस्तपरिमितपारदस्तन्निमितादेव
पानीयात्खल्वान्तरचतुर्दशगुणो गुरुर्भवति ।   अस्मा-
दिदमनुमीयते यद्भौतिकपदार्थस्योद्दिष्टावकाशे स्वगुरु-
त्वाधिक्यानुसारेण परमाणवाधिकां भवतीति ।   तत एव
स पदार्थो दृढतरावयव इत्युच्यते । अथ भौतिकपदार्थे
स्वान्तर्गतोष्णतादिक्येन विस्तृतेऽप्युष्णतानिष्काशनेन सं-
कुचितेऽपि वा तत्परमाणूनामविकृतत्वाच्चगुरुत्वं कि-
ञ्चिदपि विकृतिं नैति किन्तु तस्य घनहस्तादिमितदेशे
परमाणुसंख्याया विकृतत्वादुद्दिष्टघनाङ्गुलादिपरिमिते
तस्मिन् विकारो भवति । अच यः सापेच्चो धर्मविशेषः
परमाणुन्यूनाधिक्यात्मकः स दृढत्वसंज्ञः स्यात् ॥

द्रव्याणां जातीयगुरुत्वस्य निर्णयाय जलं मापकं
भवति यथाग्रे वक्ष्यते ॥

काठिन्यमिति । येन धर्मेण एकं द्रव्यमपरस्य पृष्ठं
छिनत्ति स धर्मः काठिन्यसंज्ञः स्यात् । अद्यावधि-
ज्ञातानां द्रव्याणां मध्ये हीर एव कठिनतमो ऽस्ति यः
सर्वाणि ज्ञातद्रव्याणि छेत्तुं शक्नोति ॥

स्थितिस्थापक इति । अन्यथाकृतस्य पुनस्ताद्द-
स्थ्यापादकः स्थितिस्थापकः ॥

भङ्गुरत्वेति । येन धर्मेण कठिनमपि द्रव्यं खल्पेना-
प्याघातेन भग्नं भवति स भङ्गुरत्वसंज्ञः स्यात् । यथा ।
काचः लोहपृष्ठं छिनत्ति अतः स लोहमपेच्य कठि-
नतरः तथापि काचः खल्पाघातेन भग्नो भवति ॥

ताडनवर्द्धनीयतेति । येन धर्मेण सुवर्णादि ताडि-
तं पचरूपं गृह्णाति स धर्मः ताडनवर्द्धनीयतासंज्ञः
स्यात् ॥

प्रसरच्चमतेति । यद्धर्मद्वारा सुवर्णादि तारूरूपं
च्चमते स धर्मः प्रसरच्चमतासंज्ञः स्यात् ॥

स्नेहाकर्षणमिति । महाभारं द्रव्यं सूच्च्मलोहसूच-

ज

मवलम्ब्य स्थातुं शक्नोति परन्तु अल्पभारं केवलं सू-
च्म्सीसकसूचमवलम्ब्य स्थातुं कल्पते । तथाच यः
पदार्थविशेषे विशेषमासादयन् धर्मः पदार्थस्य उ-
भयत आकर्षणे प्रतिबन्धको भवति स स्नेहाकर्षण-
संज्ञः स्यात् ॥

अथ परमाणुनिर्मितकार्याणि कदाचित् स्थिराणि
तिष्ठन्ति कदाचिच्च गच्छन्ति । अतो गमनधर्मविचा-
रः क्रियते ॥

द्रव्याणां स्थानविकारो गतिः ॥ ४६ ॥

साच सरलवक्रादिभेदादनेकधा । अधः पततो द्र-
व्यस्य मार्गे या गतिर्लोक्यते सा सरला । तिर्यक् प्र-
च्िप्तस्य द्रव्यस्य मार्गे या सा वक्रा । भुवि पततः प्र-
स्तरादेर्या गतिः सा वर्द्धमानाख्या । उच्च्िप्तस्य प्र-
स्तरादेरधोगमनारम्भकालात्याग् या गतिः सा ह्री-
यमानाख्या ॥

अथ गति: कस्मादुत्पद्यते कस्माच्च नश्यतीत्यु-
च्यते ॥

येन गतिर्जायते येन वा ह्रास-
मेति तद्वलसंज्ञं स्यात् ॥ ४७ ॥

कुलालचक्रं भ्रमणारम्भकाले स्वभ्रामकबलविरोधि
भवति तत: स्वस्थैर्यस्यापि रोधकं भवतीत्युक्तं प्राक् ।
तथैव अकस्माद्रथस्य गमनमनु तत्रस्थो मनुष्य: पश्चा-
त्पतति तथाच शीघ्रं गच्छतो रथस्य स्थैर्यमनु तत्र ति-
ष्ठन् मनुष्यो ऽग्रे पतति । तथैवच नद्यां गच्छन्ती नौ-
र्यदाकस्मात्तीरे लगति तदा तत: शीघ्रमवरोढुमुन्ति-
ष्ठन्तो ज्ञानिनस्तीरे पतन्तीति प्रसिद्धम् ॥

अथ द्रव्याणां स्थिरता स्वाभाविकीस्ति अथवा तेषां
गतिरित्यचोच्यते ॥

द्रव्यस्य स्थिरता समानसरलग-
तिश्च सदातने भवत: ॥ ४८ ॥

३६

अयमर्थः । नहि किञ्चन द्रव्यमुपाधिमन्तरा गति-
मञ्जवति नच गतिं प्राप्तं तां त्यजति नच भिन्नां दिशं
गच्छति ॥

अयोगोले भूमौ प्रचिप्ते यदि भूः समा न स्या-
त्तदा तन्मलं भटिति विनश्येत् । यदि समा
स्यात्तर्हि गमनप्रतिरोधस्याल्पत्वात् तन्मलमधिकका-
खेन मश्येत् । अन्ते तु नश्येदेब अच प्रतिरोधाभा-
पासम्भवात् । गमनप्रतिरोधकाभावस्तु खगोल एव
तच ग्रहाणां गमनङ्कदापि विकृतं न भवति ॥

सम्मुखदिश्येव कस्मिंश्चिल्लच्ये मोचितः शरौ भूम्या-
कृष्टिहेतोरुत्तरोत्तरमधो याति परन्तु यदि भूवायुः
स्थिरः स्यात्तर्हि नहि तस्य गतिरुत्तरतो दच्छिणतो वा
भवेल्किन्तु सम्मुखदिश्येव । यदि गतेरयं धर्मो न
स्यात्तर्हि नहि कश्चन शरं निश्चयेन लक्ष्ये प्रापयितुं
शक्नुयात् ॥

सपाषाणे भिन्द्पाले भ्रामिते यदा पाषाणस्ततो नि-

गच्छति तदा चापमौर्विकातो मुक्तः षर द्रव स सरलं
व्रजति परन्तु पाषाणभ्रममण्डलस्य यस्मात्स्थानात्त-
द्विर्गमे स लच्ये लगेत् तत्स्थाननिर्णयो ऽभ्यासतो ऽपि
दुःषको ऽतो ऽनेन लच्यहननमपि दुःषकम् ॥

अस्मादिदमवगम्यते यत्किस्मिश्चिन्मण्डले भ्रमहृव्यं
स्वजडताविरुद्धबलाधीनम्भवति । उदाहरणे । यदि
दूरे कश्चन मनुजः स्वहस्तं भ्रामयन् तत्परितः कन्दु-
कश्च भ्राम्यन् दृष्येत तदा दूरत्वादिकारणात् तद्धस्त-
कन्दुकयोर्मध्ये स्थितायां रज्ज्वामदृष्यायामपि तत्र र-
ज्ज्ववष्यम्भावो ऽनुमीयते । तद्वच्चन्द्रस्यान्यग्रहाणाञ्च
दृत्तभ्रमणे जाड्यानुकूलसरलगमनप्रतिबन्धकं किञ्चन
बलान्तरमनुमीयते एकेनैव बलेनासरलगमनासम्भ-
वात् ॥

अथैकस्मिन्नेव द्रव्ये युगपत्प्रयुक्तानां द्विचादिबलानां
कार्य्याणि वर्ण्यन्ते ॥

एकस्मादेव बलादुत्पन्नं गमनं सर्वदा सरलरेखा-

यामेव भवतीत्युक्तं प्राक् । अथ पूर्वबलदिश्येव यद्-
परं बलं संयुज्येत तदा गमनं शीघ्रतरमुत्पद्येत यथा
यदा गङ्गाप्रवाहे वहन्तीं नावं प्रवाहानुलोमगतिर्वा-
युर्हन्ति तदा सा नौरधिकं व्रजति । पूर्वबलविरुद्ध-
दिग्भि तद्बलेन सममन्यद्बलं संयुज्येत चेत्तर्हि गमनं न-
स्येत् यथा यदा समबलौ ऋषभौ मिथः शिरः संयो-
गं कृत्वा समेन बलेन यत्नं कुरुतः तदा उभावपि
न परावर्त्तेते । विरुद्धदिग्भि तद्बलाद्यूनाधिकमन्यद्बलं
संयुज्येत चेत्तर्हि तद्बलद्वयभेदकृतमैध्येन अधिकबल-
दिग्भि गमनमुत्पद्येत यथा यदा नौरनुकूलवायुबलेन
गङ्गाप्रवाहविलोमदिग्भि गच्छति तदा प्रवाहवायुबल-
भेदेनाधिकबलदिग्भि तद्गमनं भवति । एतत्स्पष्ट-
तरम् ॥

गमनकारणयोर्दिश्यौ यदि न समाने
नच विरुद्धे तदा गमनन्तद्दिशोर्मध्ये
भवेत् ॥ ४९ ॥

यथा नदीप्रवाहक्षेपणिभ्यां नौर्मध्यदिशि गच्छति ।
अत एव पारे इष्टघट्टे नौप्रापनार्थं नाविकः स्वीयनदी-
प्रवाहयोर्बले विचिन्त्यार्वाक्तीरात्तत्स्थानान्नावं मोच-
यति यत्सम्मुखदिशः प्रवाहदिशश्च मध्ये इष्टघट्टस्य
दिक् वर्त्तते ॥

यत्र गतेर्द्वौ हेतू भवतो ययोरेकेन सरल-
रेखायाङ्गमनम्भवेदपरेणच सदा कञ्चन
निर्दिष्टबिन्दुम्प्रति तत्र चक्राकारभ्रमणमु-
त्पद्यते ॥ ५० ॥

यथा भिन्दपाले हस्तेन भ्रामिते तत्रस्थः पाषाणो
दूरमपसर्त्तुमिच्छुरपि हस्ताकृष्ठ्वा चक्राकारमार्गे ग-
च्छति । तत्र यदि भिन्दपालः श्रीघ्रभ्रमणकाले भग्नो
हस्ताच्च्युतो वा स्यात्तदा तत्रत्यः पाषाण एकेनैव बलेन
प्रचोदितो हस्तेन प्रचिप्त इव गच्छेत् ॥

एवं तैलिकयन्त्रे ऽपि वृषभः प्रचोदितः स्वयं सरल-

मार्गे जिगमिषन्नपि कीलेबद्धतिर्यक्काष्ठावरोधेन चक्रा-
कारमार्गे मथकाष्ठम्परितो भ्रमति । यदा स कील-
बद्धकाष्ठादिमुक्तो भवेत्तदा प्रेरयितुः सकाशात् पला-
यन् गच्छेत् यतस्तत्र चक्राकारगमनहेतोर्मध्ये आदि-
मोऽयमपि अभूत् ॥

अथ तद्वलसंज्ञा उच्यन्ते ॥

येनाकृष्टिबलेन तत्केन्द्ररूपनिर्दिष्टबिन्द-
भिमुखमाकृष्यते तत्केन्द्राकृष्टिबलसंज्ञं स्या-
दथच येन बलेन तद्विन्दोः दूरं सरलरेखा-
यामपसरणशीलं स्यात् तत्केन्द्रोत्सृतिबल-
संज्ञं स्यात् केन्द्राकृष्टिकेन्द्रोत्सृतिबलद्वयं
केन्द्रसम्बन्धिबलसंज्ञं स्यात् ॥ ५१ ॥

केन्द्रोत्सृतिकार्योदाहरणानि उच्यन्ते ॥
तिर्यग्भ्राम्यमाणेषु कुलालचक्रादिषु स्थिताः पदार्था
दूरमपसरन्ति ॥

पेषयन्ते चित्रं धान्यं तदुपरितनभित्तया भ्राम्य-
माणं सत् तन्मध्यात्तावद्दूरतो ऽपसरति यावद्वृद्धिः
पतेत् ॥

जलपूर्णघटस्य ग्रीवां रज्ज्वा बद्ध्वा ऊर्ध्वाधरं घूर्ण-
येत्तदा केन्द्रोत्सृतिबलेन तज्जलं न पतेत् ॥

कश्चन मनुजो ऽश्वो वा धावन् यदा ऽतिभैश्च्येण
निजगमनदिग्भिन्नां दिशं गच्छति खपातकस्य केन्द्रो-
त्सृष्टिबलस्य प्रतीकाराय स खबुद्ध्यैव तद्विनदिशि न-
म्रो भवति । अचेतनशकटस्यतु झटिति गमनदि-
ग्वैपरीत्ये स बहुधा परिवर्त्तते ॥

अतिकोमलमृत्पिण्डं कुलालचक्रमध्ये स्थापयित्वा
तदतितूर्णं घूर्णयेत्तदा स पिण्डः केन्द्रोत्सृतिबलेन प्र-
सरिष्यति । एवं मृत्पिण्डवत् खाचम्परितो ऽतिजवे-
न भ्रमन्ती पृथ्वी निरच्चदेशे विस्तृताभूत् येन निरच्च-
भूव्यासस्तदच्चमपेक्ष्य किञ्चित्सान्तरैर्दशकोशैरधिको जा-
तः । निरच्चदेशे तदाधिकयवर्त्तनात्पृथ्वीचन्द्रसूर्या-

८२

ग्रामन्योन्याकृष्टिफलं किञ्चिद्विकृतमभूत् । अतो ऽय-
नचलनाद्या विकारा उत्पद्यन्ते यथाग्रे वच्यन्ते ॥

यस्तुल्यकालाभ्योस्तुल्यप्रदेशावतिक्रामति तस्य गम-
नं समानमस्तीत्युच्यते गमनमसमानमपि सम्भवति ।
तच सूचम् ॥

_____

विषमं गमनं द्विविधम्भवति अपची
यमानं वर्द्धमानञ्चेति ॥ ५२ ॥

_____

यदा पूर्वबलविरोधि बलं सर्वदा प्रतिबन्धकं भूत्वा
क्रमशः शैघ्र्यं न्यूनीकरोति तदा तद्गमनं अपचीयमानं
संचं स्यात् । यथाच । उत्क्षिप्तपाषाणस्य गमनम्पृथिव्या-
कर्षणहेतोः क्रमेण मन्दतरम्भूत्वा विनश्यति । विनष्टे
तस्मिन् पाषाणो ऽधो गन्तुमारभते । तचाधोगमनप्रकार
ऊर्ध्वगमनप्रकारेण विरुद्धो ऽस्तीति तच वर्द्धमानं गमनं
ञेयं यतश्च तत्पाषाणः केनचित्कालेन किञ्चित्प्रदेशं ग-
त्वा काञ्चन शीघ्रतां प्राप्य तेनैव कालेन तावन्तमेव
प्रदेशं गच्छेत् जडत्वात् पुनः पृथिव्याकर्षणतस्तद्गमन-

क्रमेण शीघ्रतरम्भवेत् ।   यतो यदा पाषाण: पतति
तदा पतनकाले कञ्चन वेगं प्राप्त: सन् बलान्तरसंयो-
गाभावे खजडत्वात् समानकालखण्डेषु समानप्रदेशा-
नतिक्रामेत् परन्तु तचानवरतं भूम्याकर्षणयोगो भवति
अतस्तद्वेग उत्तरोत्तरं वर्द्धते ॥

यदा कश्चन बाल: खहस्ताल्कन्दुकं भूमौ पातयति त-
दा सो ऽव्यवहितोत्तरकाले तं कन्दुकं पुन: करेण
धर्त्तुं शक्रोति परन्तु यदि तन्न्क्षणे किञ्चिद्विलम्व:
स्यात्तर्घ्यनन्तरं हस्तस्य कन्दुकानुसरणं द्यथा भवेत् ॥

उन्नताया द्दच्चशाखाया: सकाशात्पतद्दिल्व फलं कि-
ञ्चित्कालं नेचेण स्फुटं गोलरूपे द्रष्टुं शक्यते ।   ते-
नैव तदवरोहणट्ट्विक्रमो ऽपि ज्ञातुं शक्यते किन्तु
तदनन्तरं अन्तिमकाले तत्फलस्य पतनशैघ्य्रात् केवलं
रेखेव भासते ॥

द्दच्चङ्गाण्डाक्तिां्त्सिद्ध्रव्रद्रव्ये पातिते तत्पतत्ग्रवाह्दा-
कार उपरितनप्रदेशादधोभागे खावयववेगाधिक्यानुसा-

रथ उत्तरोत्तरं तनुर्भवति । यथा अनत्युच्चप्रदेशादपि गुडपाके पातिते स पाचाब्रि:सरणकाले स्थूलाकारो मन्दगामी भवति । तत: तलप्रापनात्प्रागेव स तनुरूपो जायते परन्तु स खसूच्ध्मीभवनानुसारेण श्रीघ्रं गच्छन् विलक्षणश्रैध्येण अधरपात्रं पूरयति ॥

अथामुकं द्रव्यमपेक्ष्यामुकस्य श्रीघ्रम्पतनं वायो: प्रतिबन्धादेव भवति । लेखनपत्रं मन्दं पतति किन्तु तदेव गुटिकीकृतं सच्छीघ्रम्पतति । सुवर्णमत्यन्तं गुर्वस्ति किन्तु सूच्ध्मपत्रीकृतन्तद्वायौ वहति द्रव ॥

मुद्रापरिमितलेखनपत्रं मुद्रोपरि स्थापयित्वा पातयेच्चदा तत्पत्रं मुद्रया समम्भुवमागच्छति यतो मुद्राया अध:स्थित्या स्थानान्तर्गतो वायु: पत्रपतने प्रतिरोधको न भवति ॥

यस्मिन्पात्रे यन्त्रेण वाय्वभावो विहितस्त्रच मुद्रापत्त्रौ समानश्रीघ्रतयाधो यात: ॥

दृक्तान्तिर्यक् प्रचित्तः पाषाणः क्रमेणाधो गच्छति पृथिव्याकृष्टः । यद्याकर्षणजा गतिः सरित्प्रवाहजग- तिरिव समाना स्यात्तर्हि पाषाण उपरिलिखितनौरिव सरलरेखायामेवेयात् । परं आकृष्टपाषाणाधोगतेर्व- र्द्धमानत्वान्तिर्यक्प्रचित्तः पाषाणः प्रतिचणमभिन्नदि- ग्भ्रङ्च्छतीति तद्गमनं वक्ररेखायाम्भवति इति प्रसिद्ध- म् । तद्वक्ररेखाया आकृतिर्नलिकाया जलसेचनेनापि दृश्यते ॥

वर्द्धमाना चीयमाणा चेति द्विविधा ऽपि गतिरान्दो- लके दृश्यते । तच सूचम् ॥

अवलम्बितं यत्किञ्चिदपि द्रव्यं अग्र-
तः पृष्ठतश्च स्खलितान्दोलने युज्यते
यत्तदान्दोलकसंज्ञं स्यात् ॥ ५३ ॥

स्थिरस्थानाद्बध्या ऽवलम्बितो गोलकः सामान्यान्दो- लको भवति ॥

सूचेणावलम्बितात्पाषाणादनेकविधा आन्दोलकध-
र्मा लच्चन्ते । इदानीमान्दोलकस्य धर्मविशेष उ-
च्यते ॥

प्रत्यान्दोलनस्य प्रदेशन्यूनाधिकत्वे
ऽपि काल: समान: ॥ ५४ ॥

एतद्दर्शनार्थं द्विपञ्चाशदङ्गुलात्मकसूचान्तावलम्बिता-
ल्पपाषाणादिरान्दोलकविशेष: क्रियतां तस्यैकवारमा-
न्दोलनं असुपादेनैव भवति तावति तदधिकं न्यूनं वा
प्रदेशं गच्छतु ॥

एतद्गुणविशिष्टत्वादान्दोलकस्य स कालमापने उप-
युज्यते । साधारणघटीयन्त्रं केवलमान्दोलक एव भ-
वति यस्मिन्नान्दोलके आन्दोलनसंख्यागणनार्थं चक्रा-
णि संयुज्यन्ते यच्च घर्षणप्रयुक्तगतिरोधस्य वातप्रति-
बन्धस्यच दूरीकरणाय कश्चन भार: स्थितिस्थापक-

विशिष्टं द्रव्यं वा युक्तं क्रियते ॥ अथ सूचयति ॥

आन्दोलकसूचदैर्घ्यविशेषात्तदान्दोल-
नकालविशेषो भवति ॥ ५५ ॥

आन्दोलकसूचं ह्रस्वं चेदान्दोलनं अल्पकालेन भव-
ति । दीर्घं चेद्बहुकालेन भवति ॥

कर्मप्रतिकर्मणी मिथस्तुल्ये
विरुद्धेच भवतः ॥ ५६ ॥

आकर्षणमुत्सारणं वान्तरेण द्रव्येषु चलनरूपं कर्म
नोत्पद्यत इति सिद्धं । तथाच नहि द्रव्यान्तरसम्ब-
न्धरहिते केवलद्रव्ये इदं सम्भवति किन्तु मिथः सम्बद्धेषु
द्विच्यादिद्रव्येष्वेवेदं वर्त्तते । तेषुच द्वयोर्द्वयोर्मध्ये यद्य-
प्येकं स्वीयइच्छाद्व्यान्तरसंयोगजइच्छाभावाद्वा ऽप-
रस्मान्मन्दगतिस्तथापि यथैकमपरमाकर्षत्युत्सारयति वा
तथैवा ऽपरमपि तदाकर्षत्युत्सारयति वेति बोध्यम् ॥

यथा यदि कश्चन मनुज एकस्यां नावि स्थित्वा ऽप-
रां नावं तद्बद्धरज्जोराकर्षणेनाकर्षति तदा ते नावौ
मिथः सामीप्यमामुतः । तच स यामाकर्षति सा त-
मायाति यास्वारूढः सा आकृष्टनावं गच्छति । अथ
यदि ते नावौ महत्त्वेन भारेणच मिथः समाने तर्हि
उभावपि समानगत्यैव चलतः । यदिच विषमे तर्हि
आकर्षके तयोर्नावोर्मध्ये यां कांचिदप्यारूढे ऽल्पैव
नौरारु व्रजति ॥

यतप्रग्राख्यान्नेयास्तुविशेषाद्यदा गोलको निःसरति
तदा गत्युत्पादकबलं यावन्नाले वर्त्तते नहि तावतो न्यू-
नेन बलेन यतप्नी पश्चात्सरति किन्तु यतप्रग्रां वर्त्ते-
मानस्य गत्युत्पादकबलस्य महापिण्डे विस्तृतीभवना
त्ततिरल्पा उत्पद्यते तूर्णं नश्यतिच ॥

यस्य द्रव्यस्य द्रव्यान्तर आघातो जायते तस्मिन् त-
द्द्रव्यान्तरस्यापि प्रत्याघातो जायते । एतावाघातप्रत्या-

घातौ समानबलौ मिथो विरुद्धौच भवतः । स्थि-
तिस्थापकविशिष्टद्रव्येषु इदं दृश्यते । सरलमार्गे च-
लन् कन्दुको यदि लम्बरूपेण भित्तौ पतेन्तदा स त-
स्मिन्नेव मार्गे पराष्टत्तो भवति । अथ यदि गमन-
मार्गे भित्तौ लम्बरूपो नस्यात्तर्हि भित्तौ संयोगस्था-
नाकृतस्य लम्बस्यैकपार्श्वे यावतान्तरेण स गच्छेत्ताव-
तैवान्तरेणापरपार्श्वे स पराष्टत्तो भवेत् । यथा । स-
मचतुष्कोणाग्रहस्यैकभुजप्रान्तात्तत्सम्मुखभुजमध्यचिह्ने प्र-
क्षिप्तः कन्दुकः पूर्वभुजस्यापरप्रान्तङ्गच्छति । गम-
नागमनरेखयोर्लम्बेन सहोत्पन्नौ कोणौ सदा मिथः
समानौ भवतः । इदं कोणद्वयसाम्यं उष्णतादेरपि
धर्मः । अच सूचम् ॥

पतनपरावर्त्तनकोणौ मिथः
समानौ ॥ ५७ ॥

तच दृष्टान्ता अग्रे दर्शनीयाः ॥

द्रव्येषु दृश्याः सर्वा गतयः पृथग्भूतानां संयुक्तानां वा परमाणूनां जाड्ये व्यवचरती आकर्षणोत्सारणे विना नोत्पद्यन्त इति पूर्वमुक्तम् । अधुना तदुक्त्याधारत एव पिण्डीभूतद्रव्येषु वर्त्तमानयोः स्थैर्यगमनयोर्विशेषा वक्ष्यन्ते ॥

अथ पिण्डीकार्याणि ॥

यस्मिन् द्रव्ये परमाणवः परस्परं तथा दृढमाकृष्टा भवन्ति यथा तस्यैकदेशे चालिते तत्सकलं द्रव्यं खावयवानां परस्परान्तरविकारमन्तरा चलितं भवति तत्पिण्डसंज्ञं स्यात् । एतद्धर्मफलमुच्यते ॥

पिण्डस्यैकदेशं चालयद्बलं सकल-
पिण्डं चालयति अथवा तमेकदेशं
पिण्डात्पृथक्करोति ॥ ५८ ॥

मृन्मयपात्रे तद्भित्त्यैकदेशं धृत्वा उत्थापिते इदमुपपद्यते यत्पात्रभारमपेच्य तत्परमाणूनां संयोगो दृ-

ढतरो ऽस्तीति । परन्तु यदि पात्रं शीघ्रमुत्थाप्येत
तदा तन्रीवैकदेश एव उत्तिष्ठति पात्रञ्चाध: पतति य-
त्तत्तदानीं तच्च जाड्यं भारश्च द्वे ऽपि परमाणुसंयोग-
विघाताय भवत: ॥

यदि सरला काष्ठादियष्टि: तुलादण्ड इव मध्यदेशे
ऽवलम्बिता स्यात्तर्हि तस्य प्रान्तौ परस्परं तुलयत: ।
अथ तत्प्रान्तयो: तुल्यभारसंयुक्तयो: सतोरपि समतो-
लनं विकृतं न भवति ।   ताट्टग्यष्टेर्मध्यदेशो गुरुत्व-
केन्द्रवाच्य: ।   इदं गुरुत्वकेन्द्रं प्रतिद्रव्यं भवतीतीत्ये-
तदथ सूचम् ॥
_____

यस्य कस्यापि पदार्थस्य संयुक्तपदा-
र्थसमूहस्य वा गुरुत्वकेन्द्रसंज्ञको बि-
न्दुरस्ति यम्परित: तदंशानां सम-
तुलता   ॥ ५८ ॥
_____

तस्मिन् बिन्दौ साधारे सति पदार्थो न पतति ।

यदि स बिन्दुर्निराधारः स्यात्तर्हि यस्मिन् पार्श्वे तेन स्थीयते तस्मिन्नेव पार्श्वे तद्द्रव्यं पतति । यथा । समानस्य काष्ठदण्डस्य गुरुत्वकेन्द्रं मध्य एव वर्त्तते अतस्तस्मिन् बिन्दौ अङ्गुल्यादिना अवलम्बिते दण्डः स्थिरो भवति अथाधारस्य यस्मिन् पार्श्वे स बिन्दुः स्यात् तस्मिन्नेव पार्श्वे स दण्डः पतेत् ॥

गुरुत्वकेन्द्राङ्गूपृष्ठे यो लम्बः स गुरुत्वलम्बसंज्ञः स्यात् । पदार्थो भूपृष्ठे यावन्तं देशमभिव्याप्य वर्त्तते तद्देशादहिस्तद्गुरुत्वलम्बो न पतेच्चेत्तद्द्रव्यं तिष्ठेत् अन्यथा पतेत् ॥

यदा अल्पनौस्था जना उत्तिष्ठन्ति तदा तत्कालस्य नाम नौसहिततदन्तर्वर्त्तिजनस्य गुरुत्वकेन्द्रमप्युत्तिष्ठति । अतस्तदाल्पतरेणापि नौस्थितिविपर्ययेण गुरुत्वलम्बो बहिः पतति । अनेन नौमज्जनप्रसङ्गे भयात्तदन्तर्वर्त्तिनो नोत्तिष्ठेयुः इति ज्ञायते ॥

गोलस्य गुरुत्वकेन्द्रं गोलकेन्द्र एव वर्त्तते । अतः

समानभूभागे यच कुचापि स्थितो गोलः स्थिरो भवति
यतः तद्भूसंयोगबिन्दावेव तद्गुरुत्वलम्बः पतति । परं
क्रमनिम्नभूभागे स्थापितस्य गोलस्य गुरुत्वलम्बो न क-
दाचन गोलभूसंयोगबिन्दौ पततीति स गोलोऽधो
गच्छति ॥

नहि द्रव्यमात्रस्य गुरुत्वकेन्द्रं तद्द्रव्य एव वर्त्तते । क-
ङ्कणादिपदार्थानां गुरुत्वकेन्द्रं तन्मध्यस्थाकाशे वर्त्तते ।
तज्ज्ञानप्रकारो ऽयम् । कङ्कणपालौ कस्मिंश्चित्स्थाने
सूचम्बद्ध्वा तेन सूचेन तत्कङ्कणं लम्बयित्वा तत्सूर्ध्वदिशि
तच्चक्रमध्ये एकं सूचम्बध्नीयात् ततः कङ्कणपालावन्य-
स्थाने एवमेवान्यत्सूचं बध्नीयात् तथाच सूचद्वयसंपातः
कङ्कणस्य गुरुत्वमध्यम्भवति यतत्तस्मिन्नेव लम्बिते क-
ङ्कणमवलम्बितम्भवति ॥

पूर्वोक्तपदार्थधर्मेभ्यो गतिधर्मेभ्यश्च मनुष्यलोकेशला-
घवाय अत्युपकारकाणि यन्चाणि निर्मातुं शक्यन्ते ॥

कुलालचक्रभ्रमिदर्शनेन वक्ष्यमाणप्रतिच्छाया युक्तिः
स्फुटमवगम्यते ॥

यस्मिन् कस्मिंश्चित्पिण्डे स्वाच्चम्परितश्चक्र-
वत्तुलादण्डवद्वा भ्रामिते तस्य सर्वेषामव-
यवानां केन्द्रादच्चाद्वा यथा दूरत्वं तथा तेषां
गतिविशेषः ॥ ६० ॥

चक्रे तत्प्रान्ते केन्द्रपरिध्योर्मध्यभागेच समानावार्द्र-
ष्टत्पिण्डौ स्थापयित्वा भ्रामिते केन्द्रपरिधिमध्यभागस्य
ष्टत्पिण्डमपेक्ष्य प्रान्तस्य ष्टत्पिण्डो द्विगुणां देशमतिक्रा-
मति ॥

अस्मिन् विषये वच्यमाणो ऽर्थो मनस्यवधार्यताम् ॥

असमवेगविशिष्टपदार्थयोः समतोलनं स-
म्भवति यदि तौ किञ्चित्कठिनद्रव्ये तथा स-
म्बद्धौ स्यातां यथा चलनकाले अल्पतरप-

दार्थस्य गमनं महत्तरस्य भारप्रमाणाधि-
कानुसारेण अधिकं स्यात् ॥ ६१ ॥

एष एव सिद्धान्तः सकलयन्त्रनिर्मितिशास्त्रस्य मूलं
भवति । अनेनैवोत्तोलनदण्डादीनि यन्त्राणि निर्मि-
तानि भवन्ति यद्वारा शिल्पी यद्बलं जगति प्राप्नोति
तदिष्टकार्ये योजयितुं शक्नोति ॥

तच्चोत्तोलनदण्डाख्यं धातुकाष्ठादिनिर्मितं समान-
स्थौल्यविशिष्टमस्ति ॥

यद्युत्तोलनदण्डमध्यं आधारे स्यात्तर्हि तस्य द्वौ भु-
जौ समानौ स्याताम् ॥

अथ तद्भुजाग्रयोः समानगुरुत्वपदार्थाभ्यां संयुक्त-
योरपि भुजद्वयसाम्याविघातः स्यात् । अनयैव युक्त्या
व्यवहारे पदार्थगुरुत्वमापनार्थं तुलायन्त्रमकारि ॥

यद्युत्तोलनदण्डमध्यमाधारे न स्यात् किन्तु दण्ड-
तृतीयांशचिह्नमाधारे स्यात्तर्हि यस्मिन्पार्श्वे दण्डस्य द्वौ

तृतीयांशौ वर्त्तेते तच गुरुत्वकेन्द्रं स्यात्तदा दीर्घभुजगु-
रुत्वमानं ह्रस्वभुजगुरुत्वमानद्वयेन तुल्यबलन्दृश्यते ॥

अतो यस्य तुलायन्त्रदण्डस्य भुजौ विषमौ स्यातां
तत्क्रयविक्रयकाले छलार्थमुपयुज्येतेति स्पष्टम् ॥

यदि दण्डस्य चतुर्थांशचिह्नमाधारे स्यात्तर्हि दी-
र्घभुजगुरुत्वमानं ह्रस्वभुजगुरुत्वमानचतुष्कादिना तुल्य-
बलं स्यात् अपिच यदि दीर्घभुजगुरुत्वङ्किञ्चिदधिकं
स्यात् तर्हि तत् ह्रस्वभुजे स्थितमन्यद् महद्गुरु उत्थाप-
येत् । अतो यदतिगुरु नानेकजनहस्तैरुत्थापयितुं
शक्यते तदुत्तोलनदण्डोपायेन एकोऽपि मनुज उत्था-
पयितुं शक्नोति ॥

यावता कालेन ह्रस्वभुजस्थन्द्रव्यम् अल्पदेशमतिक्रा-
मति तावतैव दीर्घभुजस्थम्बलं महाप्रदेशमतिक्रामति
इति प्रत्यक्षम् । यन्त्रमात्रे फलार्थं यथा यथा बलाल्प-
त्वन्तथा तथा कालाधिकम्बोध्यम् । नचैतावता यन्त्र-
मकिञ्चित्करमस्तीत्यनुमेयं यतः यत्कार्यं शरीरबल~

सहस्रेणापि असमाप्यन्तञ्चेत्कालाधिक्येन सम्पद्येत तर्हि
स महालाभो ज्ञेयः ॥

यच सरित्प्रवाहजं वातप्रवाहजं वा तप्ततोयवाष्पवि-
स्फूतिजं वा बलङ्गमने हेतुर्भूत्वोपतिष्ठेत तच मनुष्य-
स्तद्बलं यन्त्रोपायेन वशीकृत्य तेन खसेवकवत् कर्म का-
रयितुङ्कल्पते । इदं यन्त्राणामधिकम्फलं ज्ञेयम् ॥

इति अद्रवद्रव्यधर्माः ॥

अथ जलादिद्रवद्रव्याणां धर्मा उच्यन्ते ॥

द्रवपदार्था निजकणानां सङ्घट्टनाल्पत्वात् न पर्वत-
वत् राशीभवन्ति किन्तु गुरुत्वाकर्षणवशात् पतन्ति ।
एतद्धर्मकथनार्थं सूचम् ॥

द्रवपदार्था सदा समानपृष्ठा
भवितुमर्हन्ति ॥ ६२ ॥

तल्लक्षणानां गुरुत्वप्रयुक्तमधःपतनं निरपेच्चमस्तीति
सर्वतस्तेषां मिथःपीडनम्भवति । ऊर्ध्वभागे यत्तें-

षाम्पीडनं तत् गुरुत्वविरुद्धं नास्ति किन्तु अधःपीडन-
प्रयुक्तत्वत् ॥

यथा । यदा नालविशिष्टे पात्रे जलम्प्रचिप्यते तदा
किञ्चिज्जलं नालान्तरमारुह्य प्रांशुत्वेन पात्रस्थतोयसम-
म्भवति पात्रतले स्थिता जलकणा उपरिस्थितजलकणैः
पीडितास्सन्तो यत्र निःसरणमार्गः स्यात् तेन गच्छन्ति
इह निःसरणमार्गो नाल एव प्राप्यते तैः अतस्तैस्तत्र
गम्यते ॥

एवं जलधर्मं ज्ञात्वा समीपपर्वतकन्दरे जलाशयं च
विलोक्य दीर्घं लोहनालं तत्र संयोज्य तद्द्वारा तज्जलं
स्वनगरे अत्युच्छ्रितगृहाणामुपरिभागमपि नेतुं शक्यते ।
युरोपाख्यदेशवर्तिनगरेषु अयं महोपकारको जलप्राप-
स्योपायः क्रियते । अत्र तज्जलाशयस्थानेन नगरस्था-
नाद्वश्यं उच्छ्रितेन भवितव्यं यतो मूलस्थानादुच्च-
तरस्थाने नालद्वारा जलं न गच्छति यथा नालविशि-
ष्टपात्रे दृश्यते ॥

अथ यदि पात्रस्थितजले अतिसूक्ष्मो नालः स्थाप्येत
तदा तन्नालान्तर्गततोयम् उच्छ्रितत्वेन पात्रस्थतोयाद-
धिकं भवति नालपार्श्वानां आकर्षणात् । एवं दीप-
वर्तिस्थातिसूक्ष्मनालद्वारा तैलं ऊर्ध्वं गच्छति अपिच
त्वचायामतिसूक्ष्मनालैः जलमुपरि गच्छति ॥

अथ जलस्य जातीयगुरुत्वमापकत्वमुच्यते ॥

द्रव्याणां जातीयगुरुत्वस्य निर्णयाय
जलं मापकं क्रियते ॥ ६३ ॥

जातीयगुरुत्वं सदा सापेक्षं भवति यथा लोहम-
पेक्ष्य कर्करी जात्या लघुरस्ति किन्तु काष्ठमपेक्ष्य गु-
रुर्भवति ॥

तच्च गुरुत्वमेवं निर्णीयते । पदार्थस्य सामान्यरूपेण
गुरुत्वं विज्ञाय पश्चात् तज्जले स्थापयित्वैतद्गुरुत्वन्यून-
ता निश्चेतव्या तदा गणितेन तस्य जातीयगुरुत्वं झटि-
ति ज्ञायते ॥

यथा । किञ्चित्सुवर्णखण्डमेकोनविंशतिपलमितम-
स्तीति ज्ञात्वा जले स्थितं तत्खण्डमष्टादशपलमित-
म्भवति इति जानीयात् । ततस्तत्खण्डदेशे स्थितं
जलम्पलमितं स्यात् इत्यनुमाय सुवर्णस्य जातीयगुरु-
त्वं जलगुरुत्वमपेक्ष्य एकोनविंशतिगुणं स्यादिति बो-
धयेत् ॥

यस्य पदार्थस्य जातीयगुरुत्वं ज्ञातव्यं स जले चे-
प्यः तदा यदि यावत्स जले मग्नस्तथैव तिष्ठेत् नाधो
गच्छेत् तदा तस्य जातीयगुरुत्वं जलगुरुत्वेन तुल्यं
ज्ञेयं यतो यस्य जलांशस्य स्थाने स पदार्थः स्थितः स
जलांशो यावता बलेन पूर्वमवलम्बितस्तावतैव बलेन
तत्तुल्यगुरुत्व एव पदार्थोऽप्यवलम्बितःस्यात् ॥

यः कोऽपि पदार्थो जले प्रचिप्तः सन् स्वगुरुत्वपरि-
मितजलं अन्यच्च कृत्वा स तत्र तिष्ठति परं यदि तन्म-
हत्त्वं तद्गुरुत्वमितजलमहत्त्वादधिकं स्यात् । यथा ।
मृत्पिण्डः जले चिप्तः सन् मज्जेदेव यतस्तन्महत्त्वं स-

गुरुत्वमितजलमहच्छादूनं भवति परं स एव घटीकृतः
तरति यतः खगुरुत्वमितमेव जलमन्यचकरोति खगु-
रुत्वमितजलमपेच्य खमहच्चस्याधिक्यात् ॥

एवं नौः खभारानुसारेण जले किञ्चिन्मज्जति । तथा
लोहनिर्मिता नावोऽपि तरन्ति । गङ्गायां या बाष्पीय-
नावो दृश्यन्ते ता लोहनिर्मिता एव सन्ति ॥

यस्य पदार्थस्य जातीयगुरुत्वं जलगुरुत्वान्न्यूनं स्यात्
तस्य जले मज्जनाभावात्तज्जातीयगुरुत्वं कथमवगन्त-
व्यमित्येतदर्थन्तस्मिन् लघुपदार्थे ऽवगतजातीयगुरुत्वो
गुरुपदार्थस्तथा संयोज्यो यथा स लघुपदार्थो जले म-
ज्जेत् ॥

जलादीनां स्थितिस्थापकोऽत्यल्पोऽस्ति । द्रवीभूतानां
साधारणवाय्वादीनाम् महान्विद्यते । अथ वाय्वादि-
धर्मकथनमारेप्सुः सूचयति ॥

जलादिकात् वाय्वादिकस्य मुख्यो
विशेषः खस्थितिस्थापकः ॥ ६४ ॥

स्वस्थितिस्थापक इति । अतिबलेन पीडने वाय्वं-
शस्य परिमाणमत्यल्पं भवति तत्पीडनस्यापसरणमाचेतु
स वाय्वंशो भटिति यथापूर्व विस्तृतो भवति ॥

स्थितिस्थापकवच्चाद्वायुः सर्वतः पीडयति ॥

अथ वायुजलयोस्तमानो धर्म उच्यते ॥

_____

वायुर्गुरुरस्ति ॥ ६५ ॥

_____

अत्र प्रमाणान्युच्यन्ते ॥

वायोर्गुरुत्वादेव द्रवरूपा पदार्था नलद्वारा मुखेन
आक्रष्टुं शक्याः । यथा । नलस्यैकमग्रं जलान्तःकृत्वा
द्वितीयं वक्त्रे धृत्वा उरःस्थानं विस्तारयेत् तदा तत्स्था-
नं विशालं भवति तच्चत्यो वायुश्च स्थितिस्थापकविशिष्ट-
त्वाद्विस्तृतो भूत्वा लघुर्भवति अतो यावता बलेन जलं
पीडितम्भवति न तावता नालाग्रस्थितमिति तज्जलमब-
हिःपीडनेन नालान्तः उपरिगच्छति ॥

उपयुक्तयन्त्रेण पाचान्तर्गतं वायुं निष्काश्य तद्वायुरि-

क्षपाचस्य गुरुत्वं निर्णीय पुनर्वायुपूर्णस्य तत्पाचस्य गुरु-
त्वं निश्चीयते तथाच तद्गुरुत्वयोर्भेदान्तत्पाचमितवायो-
र्गुरुत्वं ज्ञायते ॥

वायोर्जातीयगुरुत्वं जलगुरुत्वस्य अष्टशततमांशमितं
भवति यतो यत्पाचस्य वायुपूर्णस्य गुरुत्वं तस्यैव वात-
रिक्तस्य गुरुत्वात् माषेणाधिकं तस्मिन्पाचे अष्टशत-
माषमितं जलन्तिष्ठतीति प्रत्यक्षम् ॥

भूमेरूर्ध्वं द्वाविंशतिक्रोशपर्यन्तं सामान्यवायुः वर्त्तते
इत्यनेकैर्हेतुभिरनुमीयते । वायोरुपरितनभागैरधः-
स्थिता भागाः पीडिता अतस्ते घनाःसन्ति उपरितनाश्च
शिथिलीभवन्ति । अत्युच्छ्रितपर्वतशिखरे वायोः शै-
थिल्यात् निःश्वासकरणं दुर्घटम्भवति ॥

वातगुरुत्वमापकसंज्ञकयन्त्रेण पर्वतस्योच्छ्रायोऽवग-
न्तुं शक्यते ॥

अथ तद्यन्त्रनिर्माणविधिः । द्विहस्ताधिकदैर्घ्यविभि-
ष्टं एकमुखखड्गाचनालम्पारदेन पूरयित्वा तन्मुखमङ्गुल्या

रुद्ध्वा तलमुपरि विधाय पाचान्तर्गतपारदे स्थापयित्वा
अङ्गुलीर्निष्कासयेत् तदा नालवर्तिपारद: किञ्चिदधो
यास्यति नालतले किञ्चित्स्थानं रिक्तम्भविष्यति यच्च वा-
युर्गन्तुं न समर्थ: पाचस्थपारदनिवारणात् । अतस्त-
त्स्थानस्य रिक्तत्वान्नहि नालवर्ति पारद: सामान्यवायु-
ना पीडितो भवति अतो नलस्थ: पारद: सामान्यवायु-
पीडितेन पाचस्थपारदेन अवलम्बितो वर्तते अतो यच
यदाच वायोर्गुरुत्वमधिकं स्यात् तच तदाच नालवर्ति-
पारदस्योच्छ्रता अधिका दृश्यते ॥

सामान्यतस्तस्योच्चता एकोनचत्वारिंशदङ्गुलपरि-
मिता भवति । यदि तद्यनुम्पर्वतशिखरेवा उच्चगृहो-
परिभागे वा नीयते तदा उपरितनवायो: शिथिलत्वात्
तत्गुरुत्वपीडनस्याल्पत्वात् नालवर्तिपारदोऽधोगच्छ-
ति । एवमनेकवारं परीक्ष्य नालवर्तिपारदस्याधरी-
भवनच्चानेन पर्वतोत्सेधस्य ज्ञानम्भवेदिति निश्चित्याऽधु-
ना तत्कर्मणि तद्यन्त्रं उपयोजितम् ॥

स नालस्थपारदो भ्रञ्भ्रानिलागमनात्पूर्वमपि किं-

चिद्धो गच्छति अत इदं भञ्झावातागमज्ञापकं यद्भम
महासागरगन्तृनाविकानामत्युपकारकम्भवति ॥

अथ यावता वातपीडनेन रस एकोनचत्वारिंशदङ्गु-
लमितदैर्घ्यविशिष्टोऽवलम्बितस्तावता यस्य जातीयगुरु-
त्वम्पारदगुरुत्वाद्ल्पम्भवति स अधिकदैर्घ्यविशिष्टोऽव-
लम्बितो भवेत् इत्यनुमीयतां । यथा । जलगुरुत्वम-
पेच्य पारदगुरु त्वं सार्द्धचयोदशगुणम्भवति यावत्तत्सं-
ख्यया तन्नालवर्तिपारदोच्च त्वमितिर्हन्यते तावत द्वा-
विंशतिहस्ता लभ्यन्ते प्रत्यचेणापि तन्नाले जलस्योच्छ्रि-
तिस्तावत्त्वेव दृष्यते ॥

अथ द्वयोरन्योन्यामिश्रणशीलयोः जलपारदादिद्र-
वपदार्थयोः तुल्यगुरुत्वंाशप्रमाणज्ञानार्थं युक्तिरुच्यते ।
अङ्कुशाकारा वक्रा काचनली ऊर्ध्वाग्रा धार्या ततस्तन्-
त्वग्रयोस्तौ पदार्थौ तावत्तथा चेप्यौ यावद्यथा तन्नली-
मध्यभाग एव तयोः संयोगो वर्तेत तदा तयोः पदार्थ-
योः ख्यातदेशांशप्रमाणे विपर्ययेण तुल्यगुरुत्वंाशमाने

न्नये । यथा । पानीयपारदयो: उक्तविधिना नल्यां
च्छिद्रयोरेकाङ्गुलदेशवर्तिपारद: सार्द्धचयोदशांगुलदेश-
वर्तिपानीयेन मध्यचिह्न एवावलम्बते । अत: पानीयगु-
रुत्वमपेक्ष्य पारदगुरुत्वं सार्द्धचयोदशगुणम्भवति ॥

अथेदं यन्त्रमधराग्रं यदा ध्रियते तदा तल्कुक्कुटनाडी-
यन्त्रम्रोच्यते इदमुपकारकं कुक्कुटनाडीयन्त्रम्माख्यारा-
चार्यैं: स्वश्रिरोमणिमिताच्चरायां यन्त्राध्याये वर्णितं ।
तथाहि । ताम्रादिधातुमयस्य शङ्कुरूपस्य वक्रीकृतस्य
नलस्य जलपूर्णस्य एकमग्रं जलभाण्डे अन्यदग्रं वहिर-
धोमुखञ्चैकहेलया यदि विमुच्यते तदा भाण्डजलं
सकलमपि नलेन वहि: चरति ॥

तद्यथा छिन्नकमलस्य कमलिनीनलस्य जलभ्रुड्डाण्डे
चिह्नस्य जलपूर्णसुषिरस्य एकमग्रं भाण्डाद्वहिरधो-
मुखं द्रुतं यदि ध्रियते तदा भाण्डजलं सकलमपि नलेन
वहियोति । इदङ्कुक्कुटनाडीयन्त्रं शिल्पिनां शरमेख-
लिनाञ्च प्रसिद्धमनेन बहवश्चमत्कारा: सिध्यन्तीति ॥

एतच्चन्त्वकर्मापि वातगुरुत्वेनैवोपपद्यते । तद्यथा ।
भाण्डस्थतद्द्वाद्वान्तर्गत जलमपेच्य बाह्याद्वान्तर्गतजल
स्य गुरुतरत्वादाह्याभागस्थं जलम्बहिः पतितुमारभते ।
तेनापराद्धे नलो रिक्तो भवति अतस्तस्मिन्प्रदेशे सामा-
न्यवायुगुरुत्वपीडितम् पाचस्थञ्जलमुत्तिष्ठत इति ॥
वायुः शब्दस्य मुख्यो वाहको भवति । अन्येऽपि प-
दार्थाः शब्दस्य वाहकाः सम्भवन्ति । अत्र सूचम् ॥

<hr>

कर्णसंयुक्तवाव्वादौ अत्यन्तश्रैश्र्येण स्पन्दन-
मुत्पद्यमानं शब्दप्रत्यच्चकारणम् ॥ ६६ ॥

<hr>

वाव्वादाविति । जलस्याधोभागेऽपि जातो घण्टा-
रवः श्रूयते ॥

एवमद्रवपदार्था अपि शब्दवाहका भवन्ति । यथा
लोहदण्डाग्रं सूचाभ्याम्बद्धा तत्सूचाग्रे कर्णसंयुक्तेच कृत्वा
तल्लोहदण्ड अपरलोहखण्डेन हन्यताम् । तदा तत्सू-
चद्वारा यच्छब्दप्रत्यच्चं जायते ताट्टशं वायुद्वारा न भव-

ति । घण्टादिक्षणत्पदार्थस्य वादनेन तच अत्यन्तशै-
थ्येन स्पन्दनमुत्पद्यते तःस्पन्दनेन चतो वायुरपि स्प-
न्दितो भूत्वा कर्णान्तादयति तदा शब्दप्रत्यचं जायते ॥

वाय्वादिरहितपाचस्था घण्टा वा-
दितापि शब्दं न करोति ॥ ६७ ॥

जलाशये पाषाणप्रच्छेपेण पतनस्थानमभितो ऽल्पस्त-
रङ्ग उत्पद्य क्रमशः प्रसरति । तस्य संलग्नतया परि-
तोवर्तिनि जलेऽन्येऽपि तरङ्गास्तथैवोत्पद्यन्ते । एवं यथा
ते तरङ्गाः क्रमशः प्रसरन्ति तद्वद्वायुवर्त्तिनोऽपि स्पन्द
नोत्पन्नास्तरङ्गाः सर्वतः प्रसरन्ति ॥

दूरस्थो रजको वस्त्रेण प्रस्तरन्तादयति तदा किञ्चि-
त्कालेन आघातशब्दः श्रूयते । तेन इदमनुमीयते ।
सूचम् ॥

प्रतिप्राणपादं शब्दः ७६१ हस्तमि-
तदेशं वायौ गच्छति ॥ ६८ ॥

शब्दगमनस्य वैचित्र्यं ज्ञात्वा शब्दहेतोर्दूरत्वं अनु-
मातुं शक्यते । यथा । यदि विद्युल्लतान्दृष्ट्वा चिभ-
त्याषपादान्तरेण मेघध्वनिः श्रूयेत तदा क्रोशचयाद्-
धिकदूरे मेघो नास्तीति अनुमीयते ॥

वायुतरङ्गपुनरागमनात्
प्रतिध्वनिर्जायते ॥ ६९ ॥

जलाशये उत्पन्नास्तरङ्गास्तीर हत्वा परावर्तन्त इति
दृश्यते । तद्वत् स्पन्दनोत्पन्ना वायुतरङ्गा अपि पर्व-
तगृहादिकं समानरूपं पदार्थं हत्वा पुनरायान्ति । त-
दा श्रातः शब्दः प्रतिध्वनिरुच्यते । अयं शब्दः पर्व-
तादागत इति कर्णेन बुध्यते यथा दर्पणे प्रतिबिम्बं दृ-
ष्ट्वा दर्पणापाश्वात्यभागे इदं वर्तत इति नेत्रेण बुध्यते ॥

तद्वत् यदि लम्बदिशि तरङ्गाणामाघातः स्यात्तदा
गमनागमनरेखयोरैक्यं भवेत् नो चेन्न कथमन्यथा प-
तनपरावर्तनकोणयोः समत्वम् ॥

यथा द्रवाद्रवपदार्थानां आकर्षणजं गुरुत्वमस्ति तथा
वायोरपि इति सिद्धं । परन्तु उत्सारणमुख्यकारणीभू-
ताया उष्णताया गुरुत्वं नास्तीत्यनुमीयते मानाभावात्
यद्यपि तस्यामनेके वाय्वादिकस्य धर्मा सन्ति ॥

अथ गुरुत्वरहितपदार्थानां उष्णतादीनां विचार: ॥

उष्णताधर्मो धर्मिण: कदापि न
पृथक् दृश्यते नच तच गुरुत्वज-
डत्वयो: प्रमाणमस्ति ॥ ७० ॥

अयोगोलस्तप्तो ऽपि गुरुतरो न भवति ॥

समीपवस्तूनि द्युष्णतां सम-
तया विभजन्ति ॥ ७१ ॥

यथा । तप्तायोगोले जलपूर्णपात्रे प्रचिप्ते सति
अयोगोल: शीतो भवति जलन्तु उष्णम्भवति ॥

यच्च उष्णताया न्यूनता तच्च शै-
त्यमिति व्यवहार: ॥ ७२ ॥

यथा । एकेन हिमालयप्रस्थादागतेन तथैकेन म-
ध्यदेशाद्गतेनचार्द्धपथसङ्गताभ्यामुभाभ्यां मिथो व्यतिसं-
वादे अहो अत्युष्णो ऽच वायुरहो अति शीतो ऽच वा-
युरिति न तच्च प्रत्यचं प्रामाणं तत्तत्पुरुषपूर्वानुभूतवा-
यूष्णता न्यूनाधिकभावानुसार्यनुभवभेदेन ॥

किञ्च । यथाच जनभेदन्नोष्णताशैत्यविसंवादस्त-
था जनैक्ये ऽपि पूर्वानुभूतोष्णशीतपदार्थधृतयो: क-
र्योरनुष्णाशीतजले युगपत्प्रवेशेचैकेन करेणातिशी-
तमिदं जलमिति एकेनच करेणात्युष्णमिदं जलमि-
तिचानुभवदर्शनाच्च न मनुष्यमाचेणोष्णतामापनं कि-
न्नूपायान्तरेणेति ज्ञेयम् ॥

उष्णतया पदार्था विस्तृता भवन्तीति य-

स्य पदार्थस्य विस्तरक्रमो मातुं शक्यते स
पदार्थ उष्णतामापक: सम्भवति ॥ ७३ ॥

___

अथ साधारणस्य उष्णतामापकस्य निर्माणविधि: ।
काचनलिकासहितकाचगोलं पारदेन जलादिना वा
पूरयेत् तर्हि यदा उष्णतावद्विर्भवेत्तदा पारदादिद्रव-
पदार्थ: क्रमेण विस्तृतो भूत्वा नलिकायामुपरि गत्वा
उष्णतावद्विक्रमं द्योतयति ॥

___

पदार्थविशेषेण तचोष्णताया
गमनमैच्यविशेष: ॥ ७४ ॥

___

तद्यथा । अङ्गुलीभिर्धृतं ज्वलत्तृणं यावदङ्गुलिनि-
कटे न ज्वलेत्तावदङ्गुलीषु नोष्णताप्राप्तिर्भवेत् । धा-
त्वादिषु त्वतिशीघ्रमेवोष्णता तद्वारा चलति । यथा ।
लोहसूचस्यैकस्मिन् भागे ऽङ्गुलीभिर्धृते ऽपरस्मिंस्वाग्नौ
निवेशिते ऽत्यल्पेनैव कालेनाङ्गुलीषूष्णताप्राप्ति: ॥

शून्यमार्गद्वाराप्युष्णता
गन्तुं शक्नोति ॥ ३५ ॥

तच किरणरूपा उच्यन्ते ॥

तस्मायोगोलकात्सर्वतः सरलरेखाभिस्तेजो गच्छति ।
पतनपरावर्त्तनकोणयोः साम्यं धर्म उष्णताया अपि
अतो यदा सरलरेखाभिस्समागता रविकिरणा मथनि-
स्ने दर्पणे पतन्ति तदा पतनपरावर्तनकोणयोः साम्यात्
मथ्यदिशि युक्तस्थले तावति दूरे विन्यस्तमन्नं पक्व-
म्भविष्यति ॥

अथ काष्ठादिकं घृष्टं उष्णां भवति । कौभे यघृष्ट-
काचनल्या और्णघृष्टलाचाया वा उष्णतया सह कि-
ञ्चिदन्यदप्याविर्भवति यत् ताट्टणी नली समीपस्थं
लघुतृणं पचं वा आकृष्य मुञ्चति ॥

महत्याश्च ताट्टङ्कुल्याः समीपे खाङ्गुल्यां नीताया-
मेकः स्फुलिङ्गस्ततो निःसरत्यल्पो ध्वनिश्च ततः श्रू-

श्रा

यते । यदा वायुश्शुष्को भवति तदा विडालपृष्ठे हस्तस्य घर्षणेन प्रोक्तकार्यमनुभूयते ॥

यो ञ्च हेतुः स एव मेघेभ्यो विद्युन्निसृत्य शब्दायत इत्यञ्च भवतीति ञ्चानिभिर्मेघाद्विद्युतमासाद्यावगतम् ॥

अथ विद्युदप्युष्णातावत्सदा समानभावेन स्थातुं यतते इतो यस्मिन् मेघे ऽधिका विद्यत्स्यात्ततः सा समीपस्थमल्पविद्युद्विशिष्टं मेघं वृच्चं मन्दिरं वा सशब्दं हठात्प्रविशति ॥

अथ यत्र पदार्थे विद्युद्वाहुल्यं स्यात्ततो विद्युद् गर्जनमन्तरेण हठात्तीच्णाग्र्यविशिष्टपदार्थान्तरं प्रविशति । अत एव कलिकत्तादिनगरस्था विद्युत्पातच्चतिनिरासाय गृहोच्छायाधिकोच्छायां तीच्णाग्रां लोहयष्टिं स्वगृहे निखनन्ति ॥

यदा पदार्थानां रसायनसम्बन्धी विकारो भवति

तदापि तत्र विद्युत्कार्यं प्रादुर्भवति । ताहशी विद्युद्-
गाख्वानिसम्सञ्चिका यतः सा गाख्वानिसञ्चकेनावि-
ष्कृता ॥

दूङ्गलख्ड्देशे तया क्रोशशतान्तरे एकेन पलेन दत्तं
ज्ञापयितुं शक्यते । तस्याः समग्रसामग्री चेदिह भार-
तवर्षे भवेत्तर्हि किंचन दत्तं कलिकत्ताख्यनगराद्ग्राख्य-
नगरे ज्ञापयित्वा ततो दिल्ल्याख्यपुर्यां ततश्च मुम्बया-
ख्यराजधान्यां एकघटिकावकाशेन ज्ञापयितुं शक्येत ॥

तस्या एव विद्युतः साहाय्येन दूरस्थे ऽपि भाण्डे
स्थितो बारूदाख्यश्रीघ्रदाह्यपदार्थो ज्वलयितुं शक्यः ।
यदा काचन महानौर्नर्द्यां मज्जति येनान्यतरणिग-
तागतावरोधः स्यात्सा महानौस्तद्विद्युत्साहाय्येन ततो
निष्कासयितुं शक्यते । एतदुपायद्वाराचापि देशे ग-
ङ्गायां निमग्ना नौकाश्चादयो ऽव्यवहितपूर्वकाले
निष्कासिताः ॥

तयैव कृतप्रवेशं लोहं लोहचुम्बको भवति ॥

प्रकाशो ऽपि गुरुत्वरहितो ऽस्ति । तं विना दृष्टि-
र्न भवति । अथ दर्शनानुशासनम् ॥

इह दर्शनानुशासने स्वप्रकाशपरप्रकाशपारदर्शकभे-
दात् पदार्थास्त्रिविधा ज्ञेया: । सूर्यदीपादय: स्वप्रका-
शा: । लोहादय: परप्रकाशा: । काचादय: पारदर्श-
का: । अत्र प्रसङ्गे पारदर्शका मध्यस्थसंज्ञा: स्यु: ॥

स्वप्रकाशपदार्थात्मकाशकिरणा निर्गत्य
सर्वत: सरलरेखाभिर्व्रजन्ति ॥ ३६ ॥

तत्र मार्गे यदि परप्रकाश: पदार्थ: स्यात्तदा तेन
प्रकाशावरोधात् तत्पृष्ठे तमो भवति । तत्र यदि
भित्र्यादि: स्यात्तर्हिच्छाया उत्पद्यते ॥

प्रायश्छाया अत्यन्तकृष्णा न भवति यतो यत्र छा-
या जायते तत्र प्रायोऽन्यपदार्थपरावृत्त: प्रकाश आग-
च्छति यथा दीपद्वयेन घटस्य च्छायाद्वयमुत्पद्यते तत्रै-
कस्मिन्दीपे नष्टे एकैव च्छाया कृष्णतरा भवति कि-

न्लु न सा अत्यन्तकृष्णा यतो गृहभित्त्यादिसमीपपदा-
र्थपराटृत्तः प्रकाशस्तत्र गच्छति ॥

अथ यत्र स्वप्रकाशः पदार्थः परप्रकाशात्पृथुतरो
वर्तते तत्र किरणमार्गस्य सरलत्वात् छाया क्रमेण न्यू-
नीभूयान्ते विनश्यति । एवं यदा कदाचित् सूर्यग्रहणे
चन्द्रच्छाया भुवं नागच्छति तदा क्वचित् चन्द्रं परि-
तो वलयाकारो रविर्दृश्यते ॥

यत्र स्वप्रकाशः परप्रकाशादल्पीभवति तत्र च्छाया
दूरत्वानुसारेणोत्तरोत्तरं वर्द्धते यथा दीपकृता मनु-
ष्यच्छाया कदाचिद्विंशतिहस्तपरिमाणा स्यात् ॥

प्रकाशो ऽतिशीघ्रं गच्छति ॥ ७७ ॥

सूर्यान्निर्गतः किरणो दण्डत्रृतीयांशकालेन भुवं
एति । अतो यदा किरणो भुवमागच्छति ततः पूर्व-
मेव दण्डत्रृतीयांशकालेन रविः किरणनिर्गमकालि-
कं संस्थानं त्यजति । अथ प्राप्तकिरणादिष्वेव दण्ड-

तृतीयांशेन अतिक्रान्ते स्थाने वयं रविं पश्यामः ॥

अथ प्रकाशकिरणगतिमितिः पूर्वं कथमवगतेति तदुच्यते । गुरोश्चत्वारश्चन्द्राः सन्तीति पूर्वमुक्तम् तेषाञ्च ग्रहणानि मुहु र्भवन्ति । तत्काला रविग्रहणवन्निश्चितेन ज्ञातुं शक्याः । अथ यदा रविगुर्वोर्मध्ये भू-र्वर्तते तदा गुरुः भूनिकटे तिष्ठति यदा च कुगुर्वोर्म-ध्ये रविरास्ते तदा गुरुः भूमेर्दूरे तिष्ठति । अथ गुरु-समीपवर्तनकाले सञ्जातानां ग्रहणानां दृग्गणितैक्य-ङ्कृत्वा तदनुसारतोऽन्यानि आगामिकालिकग्रहणानि ज्योतिर्विदो गणितेन निरणयन् । परन्तानि ग्रहणानि गुरुदूरवर्तनकाले दण्डतृतीयांशद्वयविलम्बेनापश्यन् । ततः दूरत्वात्किरणागमेऽयं विलम्बोऽभूदित्यनुमाय कि-रणागमनमानं गणितेनावागच्छन् । तद्यथा । भूमिसूर्य-योरन्तरं पञ्चचत्वारिंशन्नियुतक्रोशमितमस्तीति प्रसि-द्धम् । अतो भूगुर्वोर्गुरुनिकटवर्तनकालभवात् अ-न्तरात्तद्दूरवर्तनकालिकमन्तरं नवतिनियुतक्रोशैरधिकं

स्यात् इति स्पष्टम् । अत एव प्रकाशकिरणो दण्ड-
त्रितीयांशद्वयेन नवतिनियुतक्रोशान् गच्छति एकेना-
सुपादेनच लचक्रोशासन्नङ्गच्छतीति स्पष्टम् ॥

यदा प्रकाशकिरणाः परप्रकाशे पतन्ति
तदा ते प्रायः पराव्रजन्ति भित्तौ प्रचिन्ताः
स्थितिस्थापकविशिष्टा गोला इव ॥ ७८ ॥

अतोऽचापि पतनपरावर्तनकोणयोः साम्यम्भवति ।
अन्धकारमये गृहे यदि अतिसूच्छ्मच्छिद्रेण आगतः सू-
र्यकिरणो दर्पणीदरे लम्बरूपः पतेत्तदा एक एव कि-
रणो लक्ष्यते पतनपरावर्तनरेखयोरैक्यात् । पुनर्यदि
दर्पणस्तिर्यक् क्रियते तदा तथा न भवेत् ॥

स्वप्रकाशपदार्थादागतैः किरणैः स पदार्थो दृश्यो
भवति । अन्येतु पराष्टत्तः किरणैरेव दृश्या भवन्ति ।
दर्पणे किरणः पतन्तीति पूर्वमुक्तं । तच्च स किरणो
ऽस्माभिर्न दृश्यते किन्तु छिद्रदर्पणयोर्मध्यावकाशे रजो

बाहुल्यं वर्तते अतस्तद्रजसः पराव्त्तः प्रकाशोऽस्माकं
नेत्रयोर्लगति ॥

यदा प्रकाशकिरणः कञ्चित्पारदर्शकं त्य-
क्त्वा अन्यस्यापि पारदर्शकस्य मार्गेण गच्छेत्
तदा किरणवक्रीभवनं जायेत ॥ ७९ ॥

अत एव यदा दण्ड ऊर्ध्वाधररूपो जले वर्तते तदा
तस्य निमग्नोंऽशो ऽल्पीभूतो दृश्यते । यदा तिर्यक् व-
र्तते तदा स जलपृष्ठस्थाने भग्नो दृश्यते । इदङ्किर-
णवक्रीभवनसंज्ञं स्यात् । एवं प्रकाशः कस्यापि पार-
दर्शकस्य मार्गेण गच्छेत् तदापि किरणवक्रीभवनं
जायेत ॥

प्रकाशकिरणो वायुं त्यक्त्वा जलं विशतीति कल्प्यतां ।
तदा यदि स किरणो जलपृष्ठे लम्बरूपः पतेत् तदा
वक्रो न भवति किन्तु यदि लम्बरूपो न स्यात्तदा स
वक्रीभूय लम्बद्दिशि गच्छति ॥

पुनर्यदा किरणो घनविषयात् सूक्ष्मविषये गच्छति
तदा पूर्वोक्तरीतिविपर्ययेण स लम्बाद्दूरे वक्रीभूय
गच्छति ॥

एतत्प्रतीत्यर्थम् मृत्पात्रे मुद्रां संस्थाप्य तत्पात्रान्ता-
दृश्ये स्थाने न्यसेत् यत्र सूर्यदीपादिकिरणैः मुद्रास-
क्ता: पार्श्वांशा: प्रकाश्यन्ते किन्तु मुद्रा अदृष्टा स्यात् ।
ततस्तस्मिन्पात्रे मन्दं मन्दं जलं चिप्यताम् । तदा ते
किरणा लम्बादिक्षि वक्रीभूय तां मुद्राम्प्रकाशयेयु: ॥

अपिच पात्रे मुद्रां विन्यस्य तामवलोकयन् पश्चात्ता-
वदपसरेत् यावत् सा मुद्रा पात्रपार्श्वेण तिरोहिता
स्यात्तदा यदि कश्चन तत्पात्रे मन्दं मन्दं जलम्प्रचिपेत्
तदा सा मुद्रा दृष्टा भवेत् । यतो ये किरणा मुद्रा-
पराष्टत्ता: पूर्वं पात्रकण्ठमागम्य नेत्रस्योपरिभागे ज-
ग्मुस्तेऽधुना लम्बात् दूरत: नेत्रे पतन्ति ॥

अत एव सरित्सरस्यादीनां यत्तलं लक्ष्यते तद्वास्त-
वतलादुच्छ्रिततरम्भवति । तेन तरणविद्यायामकुशला

त

१२२

अज्ञानिनो बाला अच जलं गम्भीरं नास्तीति विचिन्त्य तच स्नानार्थङ्गत्वा विनश्यन्ति ॥

सूर्यंतारादीनाङ्किरणा: साधारणवायौ आगम्याधो वक्रीभवन्ति । तेन चितिजसमीपे यदा ग्रहादि खस्थं वर्तंते तदा तत वास्तवस्थानात् उच्चदेशे दृश्यते । नेचकिरणसंयोगकाले यस्या दिशो द्रव्यपराङत: स किरण आयाति तस्यामेव दिशि तद्द्रव्यं वर्तंत इति नेचेण बुध्यते ॥

यत्खस्थमूर्ध्वं स्वस्तिके वर्तंते तस्य किरणा वक्रा न भवन्ति । चितिजस्थस्य वक्रीभवनम्परमं विद्यते । तत उक्तस्य किरणवक्रीभवनन्यूनता स्यात् । अत: सूर्योदयसमये विम्बोर्ध्वभागकिरणवक्रीभवनमपेक्ष्याधोभागकिरणवक्रीभवनं अधिकम्भवति । अत एव यदा चाद्रेसाि्ि्हेतुभि: दायुर्घनोस्ति तदोदयकाले रविविम्बञ्झण्डवत् दृश्यते ॥

अथ ज्योतिषगणितविषये किरणवक्रीभवनभूतं सू-
र्यादिस्थानान्तरमवश्यमनुमेयं भवति ।   अतः कुजा-
टूर्ध्वम् प्रत्यंशं वक्रीभवनफलं निश्चित्य ज्योतिषसारण्यां
लिखन्ति ॥

काचमयसमानशिलायां यः किरणो लम्बरूपो न
पतति स पूर्वं काचप्रवेशकाले उक्तवत् लम्बदिशि व-
क्रीभूय ततो निर्गमकाले पुनर्लम्बाद्दूरे गच्छतीति इह
तद्वक्रीभवनम् द्विविधम्भवति ।   अच किरणगतौ स्व-
ल्पो विकारो भवति ॥

अथ गोलखण्डरूपे मध्योन्नतकाचे ये किरणाः पत-
न्ति तच तन्गोलकेन्द्रगामीकिरणो वक्रत्वन्नामोति अ-
न्येच प्रवेशकाले तद्वदिशि वक्रीभवन्ति निर्गमका-
लेच प्रोक्तकारणानुसारतः सर्वेंऽग्रे एकस्मिन् बिन्दौ
समुद्यन्ति ॥

सूर्यकिरणानां तत्समुदयस्थले पटादिद्रव्याणि ज्व-
लन्ति ॥

अथ मध्योन्नतकाचद्वारा सूक्ष्मः पदार्थो महान्दृ-
श्यते ॥

अथ तद्धेतुभूतमुच्यते । नेत्राद्दूरे यः पदार्थो वर्तते
सोऽल्पो दृश्यते यश्च निकटे स महान् दृश्यत इति प्र-
सिद्धम् । अथ नेत्राद्दृष्टतृतीयांशमितदूरदेशे स्थितः
पदार्थः स्वस्थनेत्रेणातिस्पष्टो दृश्यते । तदधिकसमी-
पे स्थितः ततो महानपि अस्पष्टो दृश्यते तत्तद्दूरे धृ-
तपुस्तकमालोक्य एतत्प्रतीतिरुत्पाद्या ॥

अथास्पष्टत्वे हेतुरुच्यते नेत्राग्रभागे स्वच्छजलनिर्मि-
तो मध्योन्नतकाचवत् नेत्रांशो वर्तते । तच्च स्वस्थे नेत्रे
किरणावक्रलं तादृशम्भवति येन हस्तदृष्टतृतीयांशान्तरि-
तप्रदेशादागताः किरणाः नेत्रान्तः स्थितचिचपटे पूर्वो-
क्तरीत्या एकत्र समुदयन्ति तेन दृष्टिस्पष्टता जायते ॥

दृग्दमनुष्यस्य नेत्रे प्रायस्तदंशजलं न्यूनम्भवति तदा
तत्कृतकिरणावक्रीभवनमपि न्यूनञ्जायते । अतो ह-
स्तदृष्टतृतीयांशान्तरितदेशे स्थितस्य ग्रन्थस्य वर्णाः दृग्दमनु-

जेन स्पष्टं नेश्यंते हस्तमितदूरात्स्फुटतरमवलोक्यन्ते ॥

तत् हस्तद्वतीयांशदूरात् स्फाटमालोकयितुं काच-
दर्यानिर्मितं अधिककिरणवक्रत्वकारकं यन्त्रं क्रियते ॥

अथ अत्यन्तकिरणवक्रत्वकारकमध्योन्नतकाचद्वारा
नेत्रस्यातिनिकटेऽपि स्थितः सूच्छ्मः पदार्थः स्फुटमति-
महान् दृश्यते अतस्ताटृशः काचखण्डः सूच्छ्मदर्शक इ-
त्युच्यते ॥

अथ सूच्छ्मदर्शकनिर्माणस्य सुगमोपायोऽयम् । अ-
स्थूले ताम्रादिपात्रे सूच्छ्मं इत्तच्छिद्रं क्रियताम् । तत्र
स्थितो जलबिन्दुर्गोलीभूय सूच्छ्मदर्शको भवति ॥

मध्यनिम्नकाचखण्डस्य धर्माः पूर्वोक्तविरुद्धा ज्ञेयाः ।
तद्द्वारा पदार्था अल्पतरा दृश्यन्ते ॥

कदाचित् कस्यचिद्बालकस्य चच्चुषि पूर्वोक्तस्वच्छज-
र्लानिर्मितांशस्य जलाधिक्येन किरणवक्रत्वं ताटृशम्भवति
येन किरणानां समागमो नेत्रस्य मध्ये भवति । तत्र
किरणा एकीभूयाग्रे स्वस्वरेखायाङ्गत्वा पृथक् चित्रपट्टे

पतन्ति अतस्तच दृष्टिस्पष्टत्वं न भवति । स बालको
ऽतिसमीपस्थानेच पदार्थान्स्पष्टम्पश्यति । पठनकाले
नासिकाग्रभागे पुस्तकन्दधाति ॥

अथ तस्य हस्तत्रितीयांशदूरात्स्पष्टतया आलोकनार्थं
मध्यनिम्नकाचद्वयेन ताटृशं यन्त्रं क्रियते यद्द्वारा विरु-
द्ववक्रत्वमापन्नाः किरणाः तस्य नेत्रे गच्छेयुः ॥

तस्य बालकस्य दृग्गोले यदि नेत्रवर्तिजलस्य न्यूनता
स्यात् तदा स जातु यन्त्रं विनैव स्पष्टन्द्रक्ष्यति इति
तस्य लाभः ॥

ये पदार्थाः वस्तुतः सूच्च्मा न सन्ति किन्तु दूरत्व-
हेतोः सूच्च्मा दृश्यन्ते तेषां स्पष्टत्वेन दिट्ट्चा चेत् तर्हि
तच दूरदर्शकयन्त्रम्म्प्रयोजयेत् ॥

एतद्यन्त्रस्य अनेके प्रकारा भवन्ति । तच्चैको वर्ण्यते ।
तथाहि । उपयुक्तनलिकाया अन्ते ताट्टृशो विपुलदर्श-
कः काचखण्डः स्थाप्यते येन दूरस्थितपदार्थादागताः
किरणाः वक्रीभवेयुः । अपिच किरणानामेकीभवन-

स्थानात्पूर्व नलिकाया अपरान्ते स्थापितेन मध्यनिम्न-
काचखण्डेन तङ्कलस्य ताट्य: प्रतिरोध: क्रियते येन
दृष्टिस्पष्टत्वम् भवेत् ।   तथाच तेन स पदार्थो महा-
कारो दृश्यते आसन्न इव बुध्यतेच ॥

शुक्ला:   किरणो   विचिचरूप-
किरणनिर्मितो ऽस्ति ॥ ८० ॥

अस्य प्रमाणमुच्यते ।   काचमयं गोलरूपञ्जल पूर्यां-
म्पाचं स्वगिरोऽपेक्ष्योच्चतरे स्थाने विन्यस्य हस्तेन धृत्वा-
धा सूर्यपादयोर्मध्ये तिष्ठेच्चेतर्हि ये ये विचिचा वर्णा द्-
न्द्रधनुषि दृश्यन्ते तान् सर्वान् तत्पाचे पश्यति ।   तत:
पाचे उत्थापिते ऽध: कृते वा क्रमेण पीतहरितनीलादि-
रूपाणि दृश्यन्ते ।   अथ स्वच्छजलादेतद्रूपवैचिच्यं कथ-
मुत्पद्यत इत्येतदर्थमुच्यते शुक्ला: किरणस्तद्विचिचरूप-
किरणनिर्मितोऽस्तीति ॥

परप्रकाशपदार्था: पराङ्चै: किरणैरेव दृश्यन्त इति

पूर्वमुक्तम् । परं नहि सकलपदार्थेभ्यः सकलकिरणाः
परात्रजन्ति । यस्मात् सर्वे परात्रजन्ति स एव शुक्लो
दृश्यते । यस्माळ्कोऽपि किरणो न परात्रजति स कृष्णो
दृश्यते ॥

तेजसा सह तच्र्या उष्णतापि परात्रजति प्रविश-
तिवा । अतः कृष्णवस्त्रमुष्णम्भवति शुक्लञ्च शीतम् ।
अपिच विपुलदर्शककाचखण्डकृष्णां वस्त्रं शीघ्रं दहति
न तथा शुक्लम् ॥

यस्मात्पदार्थाद्रक्तकिरणा एव परात्रजन्ति स रक्तो
दृश्यते यस्माच्च पीतकिरणाः स पीत इत्यादिरूपधर्मो
बोध्यः ॥

अथ शुक्लाकिरणवर्तिरक्तपीतनीलरूपाणां वक्रीभ-
वनशीलं समानं न भवति । पीतमपेच्य नीलस्य व-
क्रत्वं अधिकम्बोध्यं रक्तस्य च न्यूनम् ॥

एतत्प्रतीत्यर्थञ्चतुर्भुजाकृतिपार्श्वचयविशिष्टकाचख-
ण्डमन्धकारमये गृहे तथा स्थाप्यं यथा तत्र सूक्ष्माच्छिद्रे-

शान्तः सूर्यकिरणः पतत् तदा तच्छिद्रागतो यः किरणः पूर्बभित्तौ शुक्ल एव दृष्टः स रक्तपीतनीलादिरूपयुक्तो दृष्यते । चिपार्श्वकाचखण्डकृता किरणवक्रगतिः पूर्वोक्तवक्रीभवनधर्मानुसारेण विपुलदर्शकक्रृतेव वारद्वयं समानदिशि विद्यते परन्तु पार्श्वसमत्वात् न तथा किरणसमाहारो जायते ॥

यदि चिपार्श्वकाचेन भिन्ना विचिचरूपकिरणा विपुलदर्शककाचे पतन्ति तदा पूर्वोक्तवक्रीभवनधर्मानुसारेण पुनस्ते एकीभूय शुक्लां स्वरूपम्प्रकटयन्ति । एवम्प्रकारेण किरणस्य पृथक्करणम् पुनरेकीकरणाञ्चकर्तुं शक्यते ॥

कुञ्झटिकासमये रवेरागता रक्तरश्मयो नेचमागच्छन्ति । अन्येतु अधिकवक्रीभूतत्वात् नायान्तीति तत्काले रवीरक्तरूपो दृष्यते ॥

थ

अथ यथा चिपार्श्वकाचेन जलपूर्णकाचपात्रेण वा
शुक्लकिरणस्य विचित्ररूपाणि पृथक् भवन्ति तद्दर्ष-
णकाले भानुभानवः पतद्विन्दुसन्दोहे पतित्वा पृथ-
ग्भूत्वाच नेत्रमायान्ति । रक्तपीतनीलादिप्रत्यचं सू-
र्येनेचविन्दुजातकोणपरिमाणाश्रितमिति पूर्वोक्तजल-
पूर्णकाचपात्रस्य परीचणेन सिद्धम् । अथ ये जल-
बिन्दवः र्विनेत्राभ्यान्तुल्यकोणेषु वर्तन्ते ते त्तत्त्त्ति-
वर्तिन एव भवन्ति इति चेत्रगणितत्त्तेन न्नायते ।
अत इन्द्रचापाकारो त्त्त एवास्तीति ॥

जडवस्तूनां गतिर्बलकार्याणि तेजःप्रभृतीनां विवि-
धानि प्रसरणानिच रेखानियमेन प्रदेशनियमेन चो-
त्पद्यन्ते तच्च्यरेखाप्रदेश्योर्धर्माणां न्नानं गत्यादिनि-
यमन्नाने सहकारि । एवं गतिबलप्रसरणानि का-
लाश्च रेखाप्रदेशादिपरिमाणेन परिच्छिन्नाः सन्तः सं-
ख्ययापि परिच्छेत्तुं योग्या यथान्यानि सर्ववस्तूनि

इत्यतः संख्यापरिमाणादिविचारस्योत्तरप्रकरणे कारि-
ष्यमाणत्वादस्मिन् प्रकरणे इन्हैव विरम्यत इति ॥

॥ इति द्वितीयाध्यायः समाप्तः ॥